Auditory Information Processing

Harunori Ohmori

Auditory Information Processing

Springer

Harunori Ohmori
Faculty of Medicine
Emeritus Professor of Kyoto University
Kyoto, Japan

ISBN 978-981-32-9712-8 ISBN 978-981-32-9713-5 (eBook)
https://doi.org/10.1007/978-981-32-9713-5

This Springer imprint is published by the registered company Springer Nature Singapore Pte Ltd.
The registered company address is: 152 Beach Road, #21-01/04 Gateway East, Singapore 189721, Singapore

Preface

When I came back to Tokyo in 1982 after 3 years of study in the United States, I had a vague idea in mind to work on issues related to excitable membrane. However, the problem was on what types of excitable membrane and on what topics of the membrane excitability I shall work. I had at that time a technique of electrophysiology in general, particularly the patch clamp. I had some knowledge about ion channels from my graduate experiments on oocyte channels. My research in the United States was about synaptic transmission in Professor Eric Kandel's laboratory, and about ion channels in Professor Susumu Hagiwara's laboratory. I had acquaintance with several patch clampers who emerged rapidly after the success of single channel recordings by Neher and Sakmann (1976). While attending the Gordon Conference on ion channels in the summer of 1982, I thought naively that a sensory receptor cell is an interesting subject, particularly the mechanism of sensory transduction, since it is the abridgement of the nervous system. Sensory receptor cells receive external signals, transduce them to electrical signal, transmit such signals through a synapse to afferent nerve fibers as presynaptic elements, and simultaneously receive efferent signals as postsynaptic elements. Accordingly, the elements of the nervous system are condensed in a single cell, except for the higher-level integration function of neural signal. I imagined conducting experiments that start at receptor cells, making a bottom-up study to investigate step-by-step the ascending signal pathway, and studying the central processing of sensory information. The problem was what receptor I should choose as the target. Eventually, I chose the auditory receptor, the hair cell. I knew, in other fields such as vision, that sensory receptor mechanisms were already explored extensively by many people. Of course, many people were working with hair cells at that time, but these facts were not visible to my eyes because of my ignorance.

I started patch clamp experiments on isolated hair cells in Tokyo, immediately after I had built up my own rig for electrophysiology experiments, and then isolation of hair cells, following the protocol of Lam (1972) who developed procedures of enzymatic isolation of photoreceptor cells. Professor Kunitaro Takahashi, my boss, provided me a fine Nikon inverted microscope equipped with Nomarski optics, a couple of Narishige's hydraulic micro-manipulators, and a DEC microcomputer (PDP-11LSI). I built my own version of a patch clamp amplifier by soldering many electronic parts. As for the animal from which to isolate hair cells, I chose chick. At that time and even now, guinea pig is one of the standard animals to investigate issues about hearing, and very few people use chicks for the auditory research.

The reason of this choice was simple but rather a long story. I was a research assistant in the University of Tokyo, Faculty of Medicine, after returning from the United States. However, I practically had no funds to conduct my research. My laboratory occupied a small corner of Professor Kunitaro Takahashi's laboratory. Across the corridor, there was the Department of Neuroanatomy where people were working with subjects on neural development using chicken embryos. They had couple of incubators for fertilized eggs. Eggs were maintained in the incubator after tracer injections into the spinal cord of the embryo. Many fertilized eggs were incubated at one time, and some were not subjected to operation. From those eggs, very frequently healthy chicks hatched out and were not used for experiments. So, I could get chicks free of both charge and daily care. I used them to obtain hair cells.

Hair cell isolation from the chicken vestibular and cochlear organs was not difficult, although the cells yield was very low. The enzymatic isolated hair cells could settle stably on a clean glass floor of a recording chamber. Then, I conducted patch clamp experiments on them. I chose hair cells that had a hair bundle of well-organized shape. Many hair cells were injured during cell isolation procedure. Those cells appeared swollen and had disarrayed hair bundle. Among cells debris, I looked for hair cells that had a well-organized hair bundle and used them for the patch clamp experiment.

The patch electrode recording from hair cells was quite stable and was fairly resistive to mechanical disturbance that was frequently transmitted from the floor of laboratory during experiments. The recording table was isolated from the floor vibration by using several tennis balls and a sandbox at the bottom of each leg of heavy iron table. The isolation of the floor vibration was not working practically. I tried to minimize the visible vibration of electrode by fastening the heads of all hydraulic manipulators to the stage of microscope on which isolated hair cells were plated in the recording chamber. They perhaps vibrated, but the vibration was in one body and was not visible through the microscope to my eyes. I first recorded from ion channels of the hair cell membrane. Looking at the isolated hair cell under the microscope, I wondered how I could apply stimulation to the hair bundle and how I could record the transduction current. For that experiment, I had to build a micro-mechanical stimulator. I built a stimulator device using a piezoelectric plate that was in an electric buzzer. The buzzer was from an old kit for elementary electronics that were my favorite toys in my childhood.

It was really surprising to see how nature has developed such fine mechanisms of encoding auditory signals to electrical signals in hair cells and then the transmission of the signal to auditory nervous system. It was also a great pleasure for one investigator to have such chances to explore and to disclose the exact steps of auditory signal transduction and transmission. I have included in this book some history of my own research activity that utilized existing technologies in new ways or added several ideas to the ongoing experiments. Many of those experiments were performed for the first time by me and the first time even in the history of auditory research.

Kyoto, Japan
January 2019

Harunori Ohmori

References

Lam DMK (1972) Biosynthesis of acetylcholine in turtle photoreceptors. PNAS 69(7):1987–1991. PMID:4505678 PMCID:PMC426847. https://doi.org/10.1073/pnas.69.7.1987

Neher E, Sakmann B (1976) Single-channel currents recorded from membrane of denervated frog muscle fibres. Nature 260(5554):799–802. PMID:1086359 MCID: PMC1309001. https://doi.org/10.1113/jphysiol.1976.sp011442

Contents

Chapter 1
Hair Cell Mechano-electrical Transduction and Synapse Transmission

Airborne sound is transduced to electrical signal in the inner ear. There, the mechanical energy of sound vibrates the basilar membrane of cochlea, which generates a shear force between the basilar membrane and the tectorial membrane. These two membranes sandwich the organ of Corti located on the basilar membrane. The organ of Corti is constituted of hair cells and variety of supporting cells. Hair cells are arranged in one row of inner hair cells and three to four rows of outer hair cells in the mammalian cochlea. The cell has a hair bundle at the apical end. The mechanical displacement of the hair bundle induces a change of membrane potential that is called a transducer potential or receptor potential, and the process is called mechano-electrical transduction (MET).

Hair bundle is an assemble of stereocilia which changes the height in a stepwise manner. Stereocilia of similar height are interconnected by filamentous structure and form a rank (Hillman 1972; Bagger-Sjoebaeck and Wersaell 1973; Flock 1965; Wersaell et al. 1965). Between the ranks stereocilia are further connected. The most well-known connection is that between the tip of shorter stereocilia and the flank of abutting stereocilia of taller rank. The connections are called tip links (Pickles et al. 1984). Individual stereocilium is a very fine structure, and the diameter is about 0.2 μm. The height of hair bundle varies orderly in the mammalian cochlea from about 5 μm in the low-frequency region in the apex of cochlea to a few μm in the high-frequency region at the base. The number of stereocilia to form a hair bundle increases from about 50 in the low-frequency region to about 300 in the high-frequency region in a cochlear of birds (Tilney and Saunders 1983).

1.1 Experiments on Isolated Hair Cells

Isolation of hair cells and the whole-cell patch recording from them was first made by Lewis using a bullfrog saccular organ. He presented the achievement in the Neuroscience Society meeting in 1982. Lewis talked about the electrical resonance

H. Ohmori, *Auditory Information Processing*,
https://doi.org/10.1007/978-981-32-9713-5_1

property of hair cell membrane, but did not at all the transduction process (Lewis 1982; Lewis and Hudspeth 1983). I attended that meeting and was inspired by his talk and started thinking about how I could work on hair cells after coming back to Tokyo late in that year.

I isolated hair cells from cochlear and vestibular organs of the chick. I whole cell clamped the cell and soon stimulated the hair bundle mechanically using a glass rod (Fig. 1.1) and have succeeded in recording the mechanically induced current (Ohmori 1984a; Ohmori 1985). It was a bit surprise that even single MET channel currents were visible in whole-cell recorded hair cells (Fig. 1.2). Before conducting that recording, I expected single MET channel activity could be only detectable in a cell-attached patch recording from the surface membrane of stereocilium.

The device for mechanical stimulation was my hand made using a vibrating element of a buzzer that was included in a kit of elementary electronics for children. In the spring of 1983, I could manage to apply micro-mechanical stimulation to the hair bundle, and the first report of mechano-electrical transduction (MET) current was published in 1984 in PNAS (Ohmori 1984a).

1.2 MET Current

In response to a small-amplitude displacement of the hair bundle (a few μm) by triangular or trapezoidal waveforms (Fig. 1.2), the whole-cell recorded hair cell generated current that followed the waveform of stimulus with occasional steplike events (Fig. 1.2a). The cell was recorded using CsCl-EGTA-based internal medium to reduce the leakage current. The MET receptor potential was depolarizing at negative membrane potentials in a current-clamp mode and reversed the polarity at positive membrane potentials (Fig. 1.3). The current was blocked reversibly by extracellular application of aminoglycoside antibiotics or amiloride (Figs.1.4 and 1.10).

1.2.1 Steplike MET Currents in the Whole-Cell Recorded Hair Cell

MET currents were observed as steplike currents (Fig. 1.2a, b). The smallest step amplitude had a conductance of 49.7 ± 4.5 pS. This is likely an elementary MET channel conductance. However, current steps of 100 pS or even 200 pS were frequently observed from the whole-cell recorded hair cells (Fig. 1.2c). I thought these larger current steps reflected concerted activity of 50 pS conductance channels. The

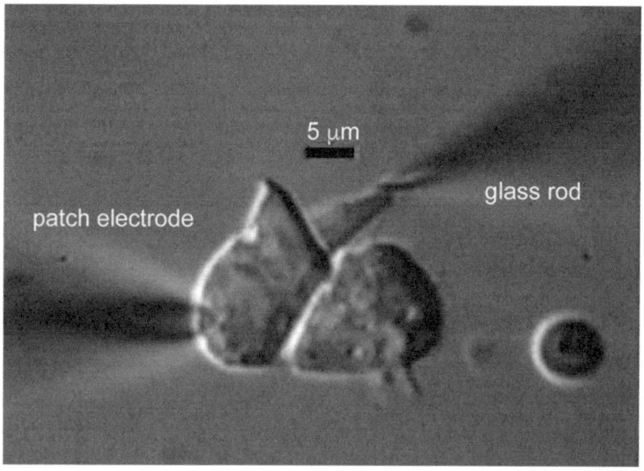

Fig. 1.1 Video image of isolated hair cell experiment
Isolated hair cell was whole cell clamped by a patch electrode, and the mechanical stimulation was applied to the hair bundle by a glass rod via piezo-electrical plate that was controlled by a computer. Hair cell was attached to the glass floor of recording chamber

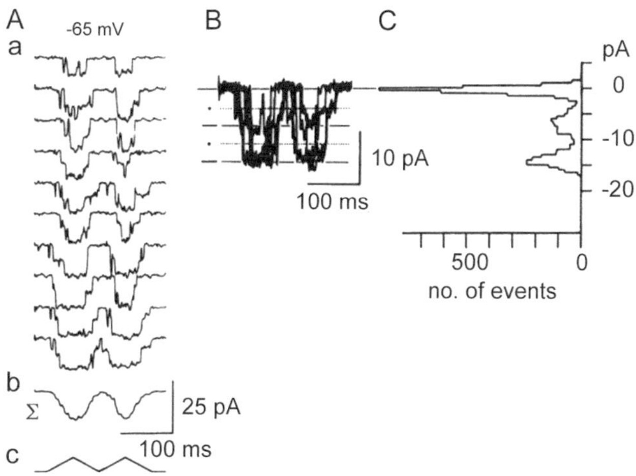

Fig. 1.2 Mechano-electrical transduction (MET) channel currents
(**A**) Steplike currents were recorded from the isolated hair cell in response to the mechanical stimulation of about 1 μm amplitude (Aa). Current traces were ensemble averaged (Ab). Hair cell was voltage clamped at −65 mV, and the mechanical stimulation was applied to the hair bundle by a glass rod (Ac). Hair cell was bathed in artificial cerebrospinal fluid (ACSF) and was recorded with CsCl-EGTA-based internal medium. (**B**) Overlaid traces from (**A**). Two large and several small current steps were detected. Large steps were indicated by solid lines and small steps by dotted lines. (**C**) Amplitude histogram of the MET currents. Peaks of currents were found nearly at multiples of 7 pA, which corresponded with the conductance of 100 pS. From the intermediate step levels in B, the smallest conductance could be 50 pS
From Ohmori (1985)

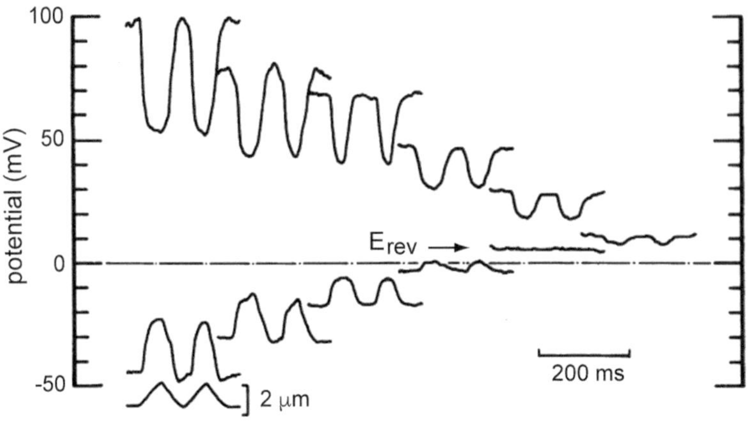

Fig. 1.3 MET potential
MET potential was induced by displacement of the hair bundle at several membrane potentials. Recording was made with CsCl-EGTA-based internal medium in artificial cerebrospinal fluid (ACSF). Optical monitor output of the glass rod motion is indicated at the bottom. MET potential reversed the polarity at +6 mV (Erev). At membrane potentials, more negative than Erev, depolarizing or positive going MET potentials were recorded, and MET potentials were negative going at membrane potentials more positive than Erev
From Ohmori (1985)

concerted activity was likely occurred because of the filamentous interconnexion of stereocilia. By mechanical stimulation the interconnected stereocilia were likely moved in a mass and generated MET currents that were integer multiples of the elementary conductance. The single-channel MET current was later recorded in isolated hair cells from several species (turtle, Crawford et al. 1991; mouse, Geleoc et al. 1997; rat, Beurg et al. 2006), and the elementary conductance reported was slightly different based on the recording conditions (Fettiplace and Kim 2014).

It should be noted here that the current through MET channel is driven by the driving force across the apical surface of hair cell between the endolymphatic potential and cell's resting potential, in a physiological condition in vivo. In mammals, it is known that the endolymphatic space has positive endolymphatic potential of up to 100 mV (80 mV in guinea pigs, Tasaki et al. 1954; Gill and Salt 1997; about 100 mV in rats, Bosher and Warren 1971; Steel and Barkway 1989). The difference between this endolymphatic potential and the resting membrane potential (−60 mV) of hair cell is the driving force. The presence of endolymphatic potential more than doubles the driving force (140–160 mV) that was built by the hair cell membrane potential alone (60 mV) and enhances the inward flow of cations through the MET channel.

The receptor potential generated by MET channel opening was measured in current-clamp experiments, and the amplitude was 24 mV at −43 mV membrane potential when using the CsCl-EGTA intracellular medium (Fig. 1.3). The MET potential change was smooth and indicated a strong low-pass filtering of the hair

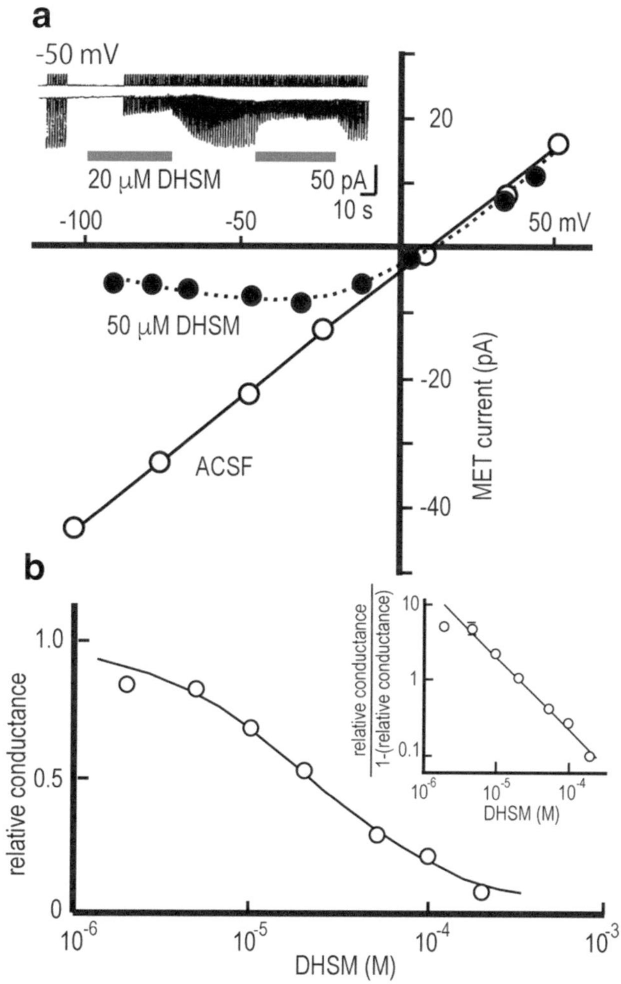

Fig. 1.4 Block of MET channel current by DHSM
(**a**) DHSM blocks the MET channel in a voltage-dependent manner. The block (filled circles) was stronger at negative membrane potentials and was reversible by wash (open circles, in ACSF). Inset shows reversible block of MET channel by DHSM (20 μM). Gray bars indicate the period of puff application of DHSM
(**b**) Dose dependence of the DHSM block measured at −50 mV. K_D was 23 μM. Inset shows the Hill's plot and the slope gives the Hill's coefficient of 0.93
From Kimitsuki and Ohmori (1993)

cell membrane. The cell's input capacitance ranged from 4.5 to 7.4 pF, and the membrane time constant, calculated for voltages between −50 and −70 mV, was in the order of 10 ms under the recording condition using CsCl-based internal medium (Ohmori 1985). However, in recordings with KCl-based internal medium, which is more physiological, the time constant should be much smaller because of the activation of Ca^{2+}-activated K^+ channel at slightly depolarized membrane and the activation of anomalous rectifier K^+ channel at the hyperpolarized membrane. The membrane time constants of 0.4–0.7 ms were reported in the guinea pig cochlea hair cells (Russell and Sellick 1978).

1.2.2 Monovalent and Divalent Cationic Permeability of MET Channel

In the isolated hair cells, the permeability of MET channel to various monovalent cations was determined from the reversal potential measurement (Ohmori 1985). Since CsCl-EGTA-based intracellular medium was used for recording, all the permeability was calculated relative to the permeability of Cs^+ (P_{Cs}), by using Goldman-Hodgkin-Katz equation (Goldman 1943; Hodgkin and Katz 1949). Even organic cations carried the current. The sequence of permeability among monovalent cations (P_X/P_{Cs}) was Li (1.39) > Na (1.22) ≥ K (1.17) ≥ Rb (1.12) ≥ Cs (1.0) > choline (0.33) > TMA (0.20) > TEA (0.17). When MET current was measured in Cs-based extracellular medium, the reversal potential was affected by the concentration of extracellular Ca ions, and the apparent Cs permeability was 0.96 with 2.5 mM Ca^{2+} and 0.99 with 20 μM Ca^{2+} (Table 1.1). Slight reduction of P_{Cs} (0.96 in 2.5 mM Ca^{2+} vs 0.99 in 20 μM Ca^{2+}) might have reflected the blocking effect of extracellular Ca ions within the MET channel, which was later reported (Fettiplace and Kim 2014).

External Ca ions were indispensable for the generation of transduction current. The minimum concentration of Ca ions for the stable generation of MET current was 20 μM in the Cs-based saline. This value was equivalent to the Ca^{2+} concentration of 23 μM or 30 μM detected in the endolymph of rat (Bosher and Warren 1978). The MET current was observed in isotonic Ca^{2+} or Sr^{2+} saline (Ohmori 1985). Sr ions could replace Ca ions without the loss of transduction activity, while generation of MET current in isotonic Ba^{2+}, Mg^{2+}, and Mn^{2+} saline needed enrichment with 1–2 mM Ca^{2+}. The permeability of divalent cations relative to the internal Cs^+ was calculated from the reversal potentials as was done for monovalent cations, and the sequence of permeability was Ca (4.65) > Sr (2.82) > Ba (2.73) > Mn (2.50) > Mg (2.41) (Table 1.1).

Small selectivity difference among alkaline cations, permeation of small organic cations, and alkaline earth divalent cations such as Ca ions and Sr ions was previously reported in the lateral line organ of the mudpuppy *Nectulus maculosus* and in the bullfrog saccule (Sand 1975; Corey and Hudspeth 1979). Sand (1975) demonstrated for the first time that Ca ions were indispensable for mechano-sensitivity of the lateral line organ of mudpuppy, and more than 1 mM was required for the maintenance of the sensitivity. He also demonstrated that Sr^{2+} could maintain the mechano-sensitivity, while Mg^{2+}, Co^{2+}, and La^{3+} suppressed the sensitivity. In the

Table 1.1 Reversal potential and relative MET channel permeability of cations to internal Cs ions

Monovalent cation	Reversal potential (mV)	P_X/P_{Cs}[a]
Li	10.7	1.39
Na	7.0	1.22
K	6.8	1.17
Rb	5.9	1.12
Cs w/2.5 mM Ca	4.3	0.96[a]
Cs w/20 μM Ca	0.3	0.99[a]
Choline	−17.0	0.33
TMA	−23.5	0.20
TEA	−25.3	0.17
Divalent cation	**Reversal potential (mV)**	**P_X/P_{Cs}[a]**
Ca	28.2	4.65
Sr	20.6	2.82
Ba	20.2	2.73
Mn	19.0	2.50
Mg	18.0	2.41

From Ohmori (1985)

Relative permeability coefficient of various cations to internal Cs ion was calculated by using the Goldman-Hodgkin-Katz equation. Reversal potentials were measured bathing hair cells in external media with 155–160 mM monovalent cation or 98–100 mM divalent cation against 160 mM CsCl-EGTA internal medium. 2.5 mM Ca^{2+} was added in solutions of monovalent cations, except for those of Cs^+. Concentrations of Ca^{2+} in Cs-based bathing media are listed. 1–2 mM Ca^{2+} was added in Ba^{2+}-, Mn^{2+}-, and Mg^{2+}-based solutions; otherwise MET current was not inducible

[a]Relative permeability ratio of external Cs^+ to internal Cs^+

sacculus of an adult bullfrog, Corey and Hudspeth (1979) studied hair cell transduction *en masse* by using the epithelium that was sandwiched between two recording chambers. They measured the transepithelial microphonic current (Corey and Hudspeth 1979). They compared the amplitude of mechanically induced current and found that all alkali cations carried the microphonic current nearly equally. The organic monovalent cation TMA carried less effectively (20% of K^+) the microphonic current. Ca ions also maintained microphonic potential, but the effectiveness was 30% of K ions. Sr^{2+} could replace Ca^{2+} activity, but Mg^{2+} and Ba^{2+} did not. They also reported that Ca^{2+} concentration of less than 10 μM was not able to maintain the mechanical sensitivity.

It should be noted that in the isolated chicken hair cells, Ca^{2+} had a larger permeability than any monovalent cations tested (Ohmori 1985). The permeability was 4.7 times larger than that of Cs^+ ions and was nearly 4 times larger than K^+ ions (Table 1.1). This Ca^{2+} permeability had significance in generation of adaptation in the hair cell transduction as later reported (Eatok et al. 1987; Kimitsuki and Ohmori 1992).

1.3 Blockers of MET Channel

1.3.1 Aminoglycoside Antibiotics

Treatment of infectious diseases with aminoglycoside antibiotics had frequently induced deafness. This is because aminoglycoside blocks the MET channel. Aminoglycoside block was studied in hair cells of many species (goldfish, Matsuura et al. 1971; lateral line organ, Kroese Van den Bercken 1982; bullfrog's sacculus, Kroese et al. 1989). Jaramillo and Hudspeth (1991) focally applied gentamicin, one of the typical aminoglycoside antibiotics, to the hair bundle of bullfrog sacculus and demonstrated a focal block of MET channel. MET channel block mechanism by dihydrostreptomycin (DHSM), another aminoglycoside antibiotics, was reported by Kimitsuki and Ohmori (1993) in detail, using isolated chick hair cells. DHSM block was reversible and was effective only at negative membrane potentials (Fig. 1.4a). Since Hill's coefficient of the block was nearly 1 (0.93) with K_D of 23 μM, it was likely that a plug by positively charged single DHSM molecule blocks the MET channel (Fig. 1.4b). Adaptation of MET current was reduced when MET current amplitude was reduced by DHSM (Fig. 1.10). Essentially the same observation was made in the experiment from epithelial preparation of bullfrog sacculus (Hacohen et al. 1989).

The block of MET channel by DHSM was membrane potential (V)-dependent, and the relative conductance (P) of the channel is formulated as follows (Kimitsuki and Ohmori 1993):

$$P(V) = 1/\left(1+[\mathrm{DHSM}]/K_D \cdot \exp(d \cdot z \cdot FV/RT)\right) \quad (1.1)$$

"d" is the fraction of membrane potential sensed by DHSM within the channel, and "z" is the valence. "F", the Faraday constant (96486.7 C/mol), "R", the gas constant (8.314 J/mol•°K), and "T", the absolute temperature (°K). Since d times z was 0.82 from the voltage dependence of P(V) and valence of DHSM molecule is 2 (z = 2), it was proposed that a DHSM molecule sensed 41% of the membrane potential when it was at the blocking site in the channel (Kimitsuki and Ohmori 1993). Later, based on the experiments using permeant blocker (FM1-43) and aminoglycoside antibiotics, free energy profile within the MET channel pore was estimated, and the blocking site was presumed to sense nearly 80% of the transmembrane potential in the MET channel (van Netten and Kros 2007). The transmembrane potential sensed by the molecule was twice of the report of Kimitsuki and Ohmori (1993) (41% vs 80%). This may be attributed to a difference in experimental procedures.

1.3.2 Amiloride

Amiloride is a blocker of epithelial Na^+ channel and suppressed reversibly the mechano-sensitivity of lateral line organ (Jorgensen 1978). Finn Jorgensen brought amiloride in my laboratory, and we tested its blocking effects on the MET channel (Jorgensen and Ohmori 1988). By applying amiloride to the bathing medium, we found a reversible block of the MET channel. The block was dose- and voltage-dependent (Fig. 1.5). The amiloride block was released at depolarized membrane potentials. The unblock potential was shifted to positive direction as the amiloride concentration was increased (Fig. 1.5). At membrane potentials more negative than −50 mV, the MET channel conductance took a limiting value, where the amiloride block had a Hill's coefficient of 1 and a dissociation constant (K_D) of 50 μM.

Slope factor of voltage dependence of the block was 15.6 mV, regardless of amiloride concentration. Based on the Eyring's rate theory, we estimated the equivalent valence of amiloride block as 1.67 (Eyring 1935; Jorgensen and Ohmori 1988), which indicated that more than one amiloride molecules participated in the block, since the valence of amiloride molecule is one. This is inconsistent with the result of Hill's constant of 1. We made a hypothesis to explain this contradiction and assumed that two amiloride molecules participated in the block of MET channel, but the change of MET current was visible only when the second amiloride molecule associated with the channel. Accordingly, the amiloride-blocking kinetics was postulated as consisted of one blocked state and two open states (Eq. 1.2). The channel was in the blocked state when two amiloride (A) molecules were bound to the receptor site (RA^2). When the receptor site was not associated with amiloride (R) or was occupied with only one amiloride molecule (RA), the channel was assumed to be in the open states (Eq. 1.2).

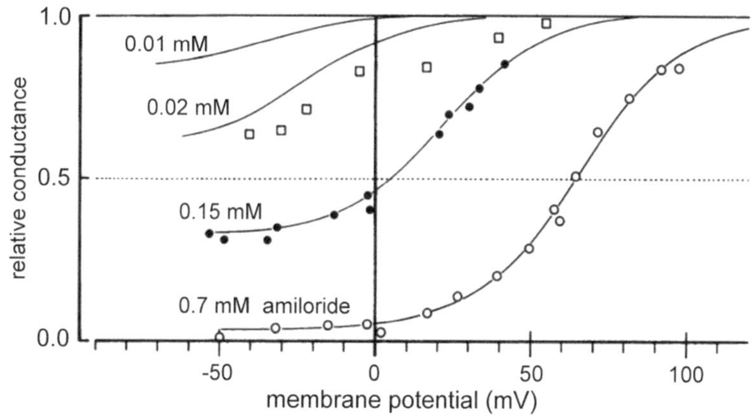

Fig. 1.5 Amiloride blocks the MET channel
Relative conductance of MET current responses was plotted against the membrane potential for several doses of amiloride. Smooth lines were calculated by Eq. (1.2)
From Jorgensen and Ohmori (1988)

$$R + 2A \underset{k_{-1}(V)}{\overset{k_1(V)}{\rightleftharpoons}} RA + A \underset{\beta}{\overset{\alpha}{\rightleftharpoons}} RA^2 \tag{1.2}$$

Two bound states (RA^2, RA) were assumed to take different binding kinetics. Transition between two open states (R and RA) was voltage-dependent with rate constants ($k_1(V)$, $k_{-1}(V)$), but the transition between the amiloride-associated open state (RA) and the blocked state (RA^2) was assumed to have voltage-independent rate constants (α, β). At positive membrane potentials, amiloride was expected to leave the receptor site ($RA \rightarrow R + A$) and makes the equilibrium between RA and RA^2 to shift toward RA ($RA^2 \rightarrow RA + A$). Accordingly, although two amiloride molecules associated with the channel, only one binding process ($RA + A \rightleftharpoons RA^2$) was visible experimentally as the change of MET current and made the Hill's coefficient apparently one (Jorgensen and Ohmori 1988). The block of MET channel by amiloride was later studied by another group (Rüsch et al. 1994), and they proposed a blocking kinetics with the Hill's coefficient of nearly 2 rather than the two open states we proposed.

It should be noted here that the most important contribution of finding amiloride block of the MET channel was to have inspired people to explore the site of MET channel by using the antibody to epithelial Na^+ channel. Amiloride was known as a blocker of epithelial Na^+ channel (Sariban-Sohraby and Benos 1986). The antibody to epithelial Na^+ channel labeled the hair bundle, and the antibody binding was blocked by amiloride and DHSM (Hackney et al. 1992; Furness et al. 1996). The immuno-gold label was demonstrated near the tip of stereocilium close to the side wall abutting taller stereocilium.

1.4 Sensitivity of the MET to Hair Bundle Displacement

Length of hair bundle is not constant along basal to apical location in the cochlear organ, and the length difference is much larger in vestibular organs, from 7 μm to more than 30 μm in the chick (Ohmori 1987). Since the length of hair bundle decreases with the increase of tuning frequency in the cochlear, the different length of hair bundle likely induces a different sensitivity to displacement stimuli (Lizard basilar papilla, Turner et al. 1981; chick, Tilney and Saunders 1983).

I sorted dissociated hair cells from vestibular organs into two classes in order to test the sensitivity difference: the one with a short hair bundle (shorter than 7.5 μm length) and the other with a tall hair bundle (taller than 12.5 μm). Relationships between the hair bundle displacement and MET current amplitude were measured and compared between these two hair cell groups (Ohmori 1987). Mechanical stimuli were applied to the hair bundle with a rigid glass rod (Fig. 1.1), and the displacement of hair bundle was measured with a resolution of 0.1 μm from the contrast-enhanced video images. The displacement applied to the hair bundle resulted in bending of stereocilia at its insertion into the cuticular plate. Displacements of hair bundle toward the taller stereocilia generated inward-going MET currents at negative membrane potentials, while displacements toward the shorter stereocilia generated outward-going MET currents (Fig. 1.6). These outward-going MET currents reflected closing of MET channels that were open at the resting position of the hair bundle. Fraction of channels open at the resting position was 0.12 ± 0.04 ($n = 7$). Therefore, the input-output relationship was not symmetrical at the resting position (Fig. 1.6). This feature was consistent with the preceding observations made in hair cells of various species or in epithelial preparation (bullfrog, Hudspeth and Corey 1977; turtle, Crawford Fettiplace 1981b; guinea pig, Russell and Sellick 1983). The displacement vs MET current relationship was not dependent on the hair cell membrane potential, and two relationships measured at −50 mV and +38 mV were superimposable after scaling (Ohmori 1987).

When the hair bundle of a shorter length (less than 7.5 μm long) was stimulated at 5 μm from the insertion to the cuticle, the minimum hair bundle displacement which could generate a detectable size of MET current was 0.01 μm (Fig. 1.7a, bottom). Displacement of 0.01 μm was not resolved by microscopy, and the displacement amplitude was estimated from the driving signal of a computer. This corresponds to 0.1 degrees of angular displacement of the hair bundle. The transduction current was linearly increased to the hair bundle displacement up to 0.6 μm (7 degrees) toward the taller stereocilia. The current amplitude was saturated with larger displacements (Fig. 1.6, filled symbols). When a hair bundle of the taller length (taller than 12.5 μm) was stimulated (toward the taller stereocilia) at 10 μm from the insertion to the cuticle, the MET current linearly increased the size of displacements up to 1.5 μm and saturated the amplitude with larger displacements (Fig. 1.6, open symbols, Fig.1.7c). When the tall hair bundle was stimulated at the height of 5 μm from the insertion to the cuticle (Fig. 1.7b), the MET current linearly followed the displacement up to 0.75 μm, but the MET current was truncated at larger displacement of 1.25 μm. The overall input-output relationship was apparently the same as that of hair cells with a short hair bundle stimulated at the height of 5 μm (Fig. 1.6, gray triangles). Apparently, a twice larger displacement was required to generate a given degree of MET current when the stimulus was applied at twice higher height (Fig. 1.6). Since hair bundle is a rigid structure, these displacement stimuli generated rotation of hair bundle at the insertion to the cuticular plate. Accordingly, the angular displacement of hair bundle is likely the factor that governs the gating of MET channel (Ohmori 1987).

Hair bundle of the ampullary organ is extremely tall. The tall hair bundle may be less sensitive to mechanical stimulation but has a wider range of operation of the transducer activity (Fig. 1.7). In the hearing organ, the hair bundle length is shorter in the high-frequency region of the cochlear than in the low-frequency region. Thus, the sensitivity range to displacement stimulus could be small in the higher-frequency

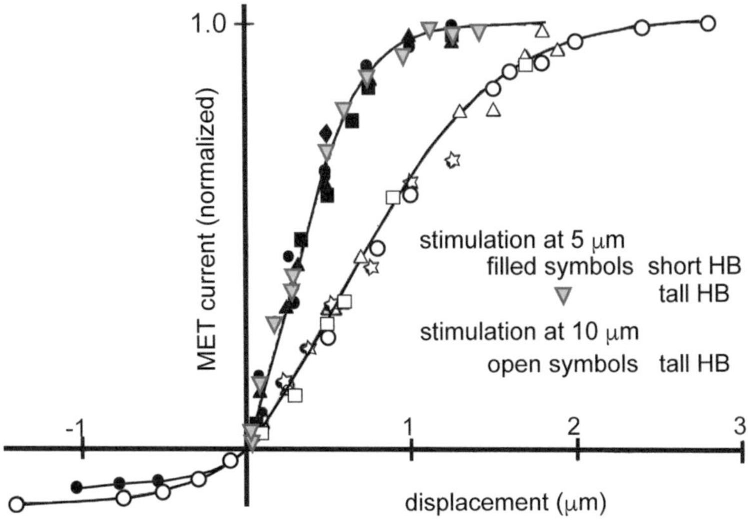

Fig. 1.6 Amplitude-displacement relationships of MET currents
Stimulus was applied to the hair bundle (HB) at two different heights from the insertion to the cuticle. Short HB was shorter than 7.5 μm, and tall HB was taller than 12.5 μm. Filled symbols are when the stimulation was applied at 5 μm to the hair bundle in both short and tall HB. Triangles pointing down (filled gray) are sampled from hair cells with a tall hair bundle. Open symbols are from hair cells where mechanical stimulation was applied at the height of 10 μm to the tall hair bundle. Note that the effective range of hair bundle displacement was twice larger, when the height of stimulation was twice higher. The hair bundle displacement toward taller stereocilia is indicated as positive and the displacement to the opposite direction as negative on the abscissa. The MET current was normalized in size, and the inward-going current was plotted as positive, and outward-going current was plotted as negative polarity
From Ohmori (1987)

region of the cochlea. Rather than the linear transduction, all-or-none type response by a short hair bundle might be more suitable to sense the high-frequency sound.

A similar dependence on the angular displacement was reported by comparison between the vestibular hair cells and cochlear outer hair cells of neonatal mice (Geleoc et al. 1997). They concluded that a rotation of the hair bundle rather than the translation determines the gating of transducer channels, independent of bundle height or origin of the cells. Some modeling reports were made to rationalize the rotation of hair bundle to the stretch of tip links (Pickles 1993) or to the shear displacement of apposition sites between adjacent stereocilia near the tip links (Furnes et al. 1997).

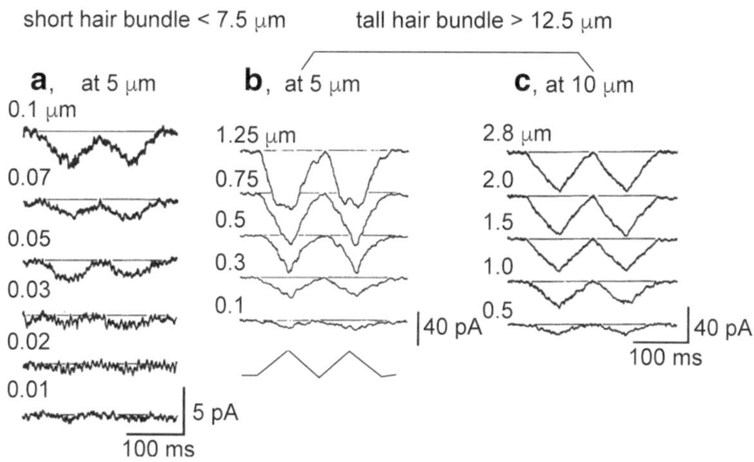

Fig. 1.7 MET currents generated by hair bundle displacement at different height
(**a**) A hair bundle shorter than 7.5 μm was stimulated at the height of 5 μm from the insertion to the cuticle. Displacement of the hair bundle as small as 0.01 μm induced the current, after ensemble averaging. The time course of hair bundle stimulation is illustrated at the bottom in **b** and is applicable also to **a** and **c**. **b** and **c** hair bundles taller than 12.5 μm were stimulated at the height of 5 μm (**b**) or at 10 μm (**c**). Note that the MET current was saturated at 1.25 μm displacement amplitude when the stimulus was applied at 5 μm (**b**), while stimulus applied at 10 μm induced smooth current even at 2.8 μm displacement (**c**)
From Ohmori (1987)

1.5 Molecular Identity of the MET Channel Is Still in Search

Hair cells are specialized to convert vibrations of a hair bundle to electrical signal via the gating of ion channels. Most of the hair cell transduction functions are well investigated as I have discussed above, except for the molecular identity of the channel. Presently, the most likely candidates of MET channel molecule are the two transmembrane channel-like proteins TMC1 and TMC2 (Kawashima et al. 2011, 2015). These channel-like proteins were found in both mouse and human from investigation of hereditary deafness genes (Kurima et al. 2002, 2003). TMC genes encode six transmembrane-spanning domains. This topology is common among proteins that have membrane-bound activity such as receptors or channels.

TMC1 mRNA is expressed in auditory and vestibular hair cells. TMC1 point mutation causes progressive hearing loss in the mouse. The mouse is known as Beethoven mouse (Bth) or deafness mouse (dn) (Vreugde et al. 2002). Point mutation introduced in *tmc1* gene reduces the calcium permeability through the MET channel (Pan et al. 2013). TMC2, together with TMC1, is expressed in auditory and vestibular hair cells. In the vestibular organs, the expression levels of both TMC1 and TMC2 were increased postnatally. TMC1 expression developed delayed to TMC2 and remained at some steady level in the adult animal. However, in the cochlear hair cells, the TMC2 expression was transient. TMC2 was changed to

TMC1 after postnatal days 2 or 3 and disappeared after postnatal day 6. The expression of TMC channels was detected by immuno-labeling near the tip of stereocilium (Kurima et al. 2015). These observations indicate that TMC1 and TMC2 are the most likely candidates of the hair cell MET channel.

However, there is one trouble to conclude that these proteins are the MET channels. The trouble is that the mechano-sensitive current is still observed in the double mutation mice of *tmc1* and *tmc2* (Beurg et al. 2016). Moreover, in these double mutation mice, the current demonstrated an opposite direction sensitivity to the hair bundle displacement: MET current was generated by the displacement of hair bundle toward the shorter stereocilia. This anomalous direction-sensitive MET current was also observed after the tip-link disruption with BAPTA or in mice of early postnatal days (Beurg et al. 2014, 2016). Accordingly, there remains argument about a possibility that TMC1 and TMC2 are not the MET channel proteins but could be proteins somehow associated with the MET channel. Therefore, we are still in search of the molecular identity of hair cell MET channel. The confusion seems to continue at least until the time when we understand how the anomalous direction sensitivity of MET current is generated.

A note must be added here about the anomalous direction-sensitive gating of MET channel. The anomalous direction-sensitive MET channel has all other properties the same as the normal MET channel has, including the permeability to calcium ions. The Ca^{2+} fluorescence increased across the apical plasma membrane when the anomalous direction-sensitive MET current was generated (Beurg et al. 2016). The anomalous direction-sensitive MET current was generated after the tip-link disruption by exposure to either BAPTA or elastase (Meyer et al. 1998). The channel was standing open. The single channel current was similar in size to that of normal polarity MET single channel. The anomalous direction-sensitive current had normal sensitivity to known blockers such as DHSM, amiloride, and gadolinium ions. These observations on anomalous direction sensitivity may indicate a possibility that the MET channel is not gated directly by the tip link.

1.6 Early Trial to Show the Site of MET

The idea that the tip link could be a gating component of MET was first proposed by Pickles in 1982, and the tip link was initially called Pickles' bundle (Pickles et al. 1984; Hudspeth 1982). The tip-link hypothesis is accepted as a most popular and realistic idea for the hair cell mechanical transduction since then. The fiber-like structure connected the tip of adjacent stereocilia of a shorter rank and the flank of a taller rank, and the location was appropriate to explain the direction sensitivity of MET channel gating. The high-resolution imaging by 2-photon fluorescence microscopy (Denk et al. 1990) succeeded to illustrate that Ca ions started to rise from the tip of stereocilia when hair bundle was displaced (Denk et al. 1995; Beurg et al. 2009). The MET channel was further argued to be localized at one side of the tip link at the tip of shorter stereocilium but not on both sides of the link

(Fettiplace and Kim 2014; Corey and Holt 2016), However, the exact role of tip link in transmitting the mechanical energy to the gate of MET channel was not demonstrated. As it was described above, the molecular identity of MET channel is not even defined yet. The difficulty to identify MET channel molecule in hair cells is likely because of the complex molecular assembly of the transduction system in the transmission of stress energy to the gate of MET channel.

Because calcium ions were the most permeable cation through the MET channel, I wondered calcium imaging might be possible to show the site of calcium influx within the hair bundle to localize the MET channel (Ohmori, 1985, 1988). Some preceding years to my experiments, calcium fluorescence imaging using fura-2 was applied to isolated smooth muscle cells and succeeded in showing graded distribution of calcium ions in cytosolic organelles (Williams et al. 1985). I applied, thus, the fluorescence imaging technology to hair cells by using fura-2 (Ohmori 1988). Confocal microscope imaging was not available at that time, and I had only an ordinary fluorescence microscope. Although the microscope was equipped with a video-intensified recording system, the spatial resolution was the level of ordinary light microscope. However, fluorescence imaging could have demonstrated calcium profiles within the hair cell and hair bundle. Furthermore, calcium profile was affected by hair bundle mechanical stimulation.

The main observation in that experiment was (1) the rise of calcium fluorescence was largest near the base of hair bundle when the hair bundle was mechanically stimulated, and (2) the extinction of the fluorescence occurred near the base of hair bundle when Mn^{2+} was included in the bathing medium (Ohmori 1988). Mn ions are permeable through the MET channel and quenched the fura-2 fluorescence. These observations raised arguments in the hair cell research community since the observation indicated that the base of hair bundle was the possible site of calcium influx or the site of MET.

In these experiments, hair cells were loaded with the esterified form of Ca^{2+}-indicator dye (Fura2-AM). Fura-2 fluorescence (510 nm) was monitored by exciting the dye at wavelengths 340 nm (F340) and 380 nm (F380). The ratio of two fluorescence images (F340/F380) indicated the intracellular Ca^{2+} level. Although hair cell was a very small structure, F340 and F380 fluorescence images and the ratio images (F340/F380) reproduced the shape of hair cell and hair bundle (Fig. 1.8). It was a little surprise for me that even the hair bundle fluorescence was visible under the conventional fluorescence microscope, even though the structure of hair bundle was very thin and the fluorescence level there was very low. When the hair cell was bathed in a high-K^+ medium (50 mM), intensity of F340 image increased while that of F380 decreased (Fig. 1.8). Because I did not have a device to change excitation filters automatically, two excitation filters (340 nm and 380 nm) were manually changed by sliding a filter bar (Ohmori 1988). However, the hair cell fluorescence still reproduced the rise of intracellular Ca^{2+} concentration by K^+ depolarization.

Fluorescence ratio was measured while applying mechanical stimulation to the hair bundle. Profiles of ratio image across hair bundle and the cell body are presented in Fig. 1.9. When the hair bundle was stimulated by a small-amplitude (about 0.2 μm) sinusoidal vibration of a glass rod (1 kHz) made at a distance, the

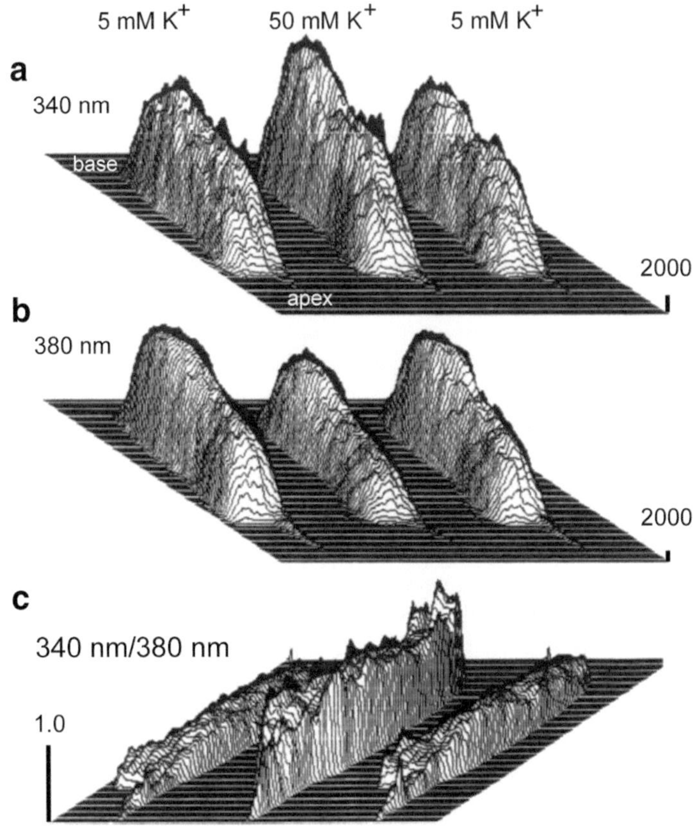

Fig. 1.8 Change of intracellular calcium concentration in hair cells
Fura-2 fluorescence indicates the calcium concentration in a hair cell after loading the cell with the membrane-permeable form of the dye (fura-2-AM). (**a**) 340 nm images and (**b**) 380 nm images when a hair cell was bathed in ACSF that had different concentrations of external KCl of 5 mM, 50 mM, and again of 5 mM in series. (**c**) The ratio images between A and B of corresponding recording conditions, showing a reversible change of Ca^{2+} concentration in the hair cell including the hair bundle. The hair bundle is pointing downward. Calibration in **a** and **b** is the fluorescence intensity of arbitrary unit
From Ohmori (1988)

fluorescence ratio was affected within the hair bundle. By stimulation applied in a Ca^{2+}-rich saline (25 mM), the fluorescence ratio was increased about the hair bundle's insertion into the cuticular plate (Fig. 1.9a profiles 2 and 5 and Fig. 1.9b profiles 5 and 7). Because Ca^{2+} influx could be made through both the MET channel and the voltage-gated Ca^{2+} channel that could be activated by the membrane depolarization induced by MET, I added Mn ion (2 mM) in the high-Ca^{2+} (23 mM) bathing medium to challenge whether Mn^{2+} was possible to indicate the influx site (Fig. 1.9b profiles 2, 4, 6, 8). Mn ions passed through the MET channel and quenched the fluorescence. Quenching was prominent near the insertion of hair

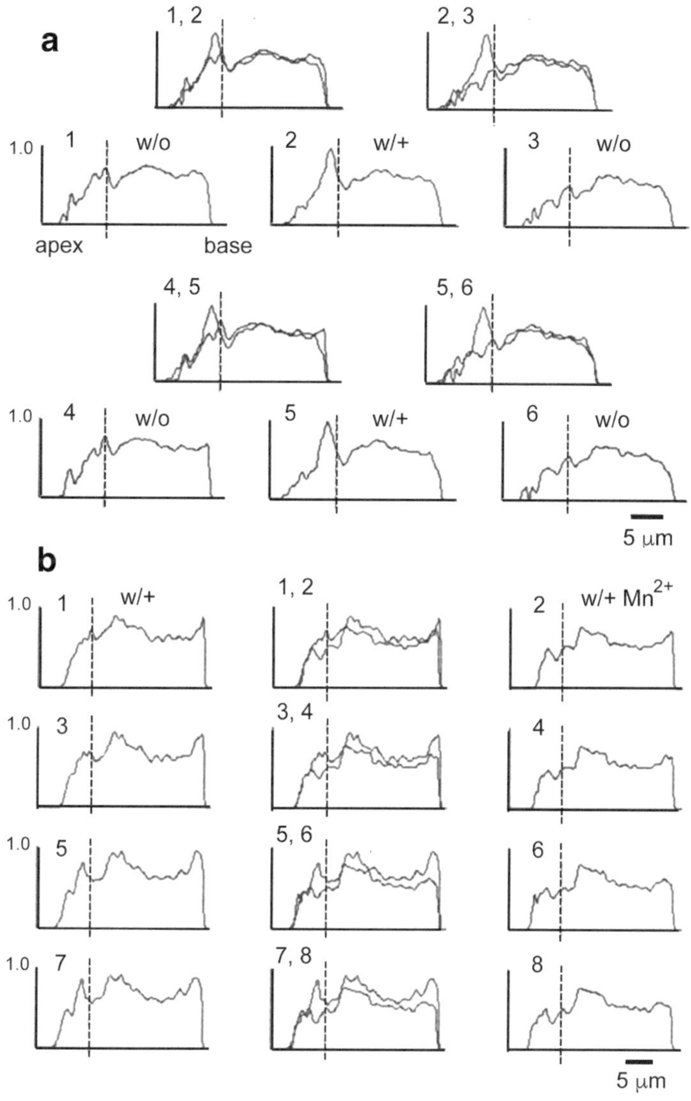

Fig. 1.9 Ca^{2+} raised and Mn^{2+} depressed the fura-2 fluorescence ratio
(**a**) Isolated hair cell was based in 25 mM Ca^{2+} saline. Small-amplitude mechanical stimulation of 1 kHz (approximately 0.2 μm amplitude) was applied to the hair bundle from a distance via water coupling by a glass rod vibration. Fluorescence measurement was made without stimulation (w/o) and with stimulation (w/+). Profiles of fluorescence ratio were measured in several parallel sections across the hair bundle and cell body. Hair bundle is pointing to the left, and the vertical broken line indicates the insertion of the hair bundle to the cuticular plate. Profiles are before the stimulus (w/o, 1, 4), during the stimulus (w/+, 2, 5), and after the stimulus (w/o, 3, 6). Two profiles with and without mechanical stimulation are overlaid in the panels (1, 2), (2, 3), (4, 5), and (5, 6). Fluorescence ratio was enhanced near the base of hair bundle during the stimulation, while profiles over the cell body were not much affected by stimulation. Fluorescence profile was always depressed in the cuticular plate

bundle to the cuticular plate when the mechanical stimulation was applied (Fig. 1.9b).

A ratio peak emerged in Ca^{2+}-rich saline near the hair bundle's insertion into the cuticular plate, while exactly at the same site, a depression of fluorescence appeared in the Mn^{2+} saline. Peak and depression appeared in symmetrical shapes (Fig. 1.9b profiles 5 and 6, and 7 and 8). Accordingly, I thought that these fluorescence ratio changes reflected the site of MET channel near the insertion of hair bundle to the cuticular plate (Ohmori 1988).

Retrospectively, results of these experiments might have been contaminated by the use of a conventional fluorescence microscope as for the imaging device and by the isolation procedure of hair cells. Firstly, the fluorescence signal could have been affected by the path length. At the base of hair bundle, the fluorescence signal was stronger because of the larger cross section of hair bundle than at the tip. Although F340/F380 ratio imaging made the image resistive to the path length difference, the quality of image data should be absolutely better from the thicker structure around the base. Accordingly, signals near the tip of hair bundle might have been missed because of the photon noise. Secondly, the taller rank of stereocilia might be injured during cell isolation more than the shorter rank stereocilia, although cells that we used had a well-integrated appearance of the hair bundle. When the hair bundle lost the integration, MET did not work, and Ca^{2+} influx did not occur by mechanical stimulation. Considering the current observations by multiphoton microscope (Denk et al. 1995; Beurg et al. 2009), which indicated an increase of Ca^{2+} fluorescence near the tip of hair bundle, some of these recording conditions or combinations of them might have affected the fluorescence calcium signal to appear stronger at around the base of hair bundle in my study.

1.7 Adaptation of MET Current

In contrast to the presence of adaptation kinetics in the microphonic current (Corey and Hudspeth 1983; Eatok et al. 1987), the whole-cell recorded MET current did not show adaptation initially in my study, even at large negative membrane potentials (Ohmori 1985, 1987). Current showed some decline during steady holding of the hair bundle at a displaced position (Ohmori 1987). The decrease of current was

Fig. 1.9 (continued) (**b**) Profiles of fluorescence ratio were measured from a hair cell, when the hair bundle was stimulated in 25 mM Ca^{2+} saline (left column, profiles 1, 3, 5, 7) or in the saline that contained 2 mM Mn^{2+} and 23 mM Ca^{2+} (right column, profiles 2, 4, 6, 8). These profiles are overlaid in the middle column. Fluorescence ratio was depressed over the cell body in Mn^{2+} saline, and the depression was marked at the base of hair bundle (profiles 2, 4, 6, 8), in contrast to the rise in 25 mM Ca^{2+} saline (profiles 3, 5, 7). Both Ca^{2+} and Mn^{2+} are permeable through the MET channel
From Ohmori (1988)

stepwise and appeared likely due to the slippage of stimulating glass rod from one rank to the other rank of stereocilia. This absence of adaptation was turned out to be due to a strong buffering capacity of Ca^{2+} by a high concentration of EGTA (5 mM) used in the intracellular medium. It was later demonstrated that enhanced Ca^{2+}-chelating capacity of BAPTA reduced adaptation (Crawford et al. 1989; Ricci and Fettiplace 1997; Corns et al. 2014).

Adaptation of MET current is an important mechanism to extend the sensitivity of hair cell transduction to large displacement of the hair bundle. The extent of adaptation was higher at higher concentration of Ca^{2+} in the bathing medium, which suggested that adaptation is controlled by the influx of Ca ions through the MET channel (Corey and Hudspeth 1983; Crawford et al. 1989, 1991). We, therefore, are interested in how the intracellular Ca^{2+} concentration affected adaptation.

By photolysis of a caged calcium compound Nitr-5, we step increased the intracellular Ca^{2+} concentration in isolated hair cells of the chick (Kimitsuki and Ohmori 1992). Photolysis of Nitr-5 by a flash of ultraviolet (UV) light induced outward currents at −50 mV (115 ± 82 pA) when a hair cell was whole-cell recorded with KCl-based intracellular medium which did not contain Ca^{2+}-chelating compound (Fig. 1.10). MET current generated at −50 mV showed a decay after step displacement of the hair bundle (Fig. 1.10b). This decay reflects the adaptation of MET current, and the kinetics of adaptation was accelerated after UV uncaging of Nitr-5 (Fig. 1.10c). The adaptation was further accelerated by hyperpolarization of the membrane but was eliminated in 20–100 μM Ca^{2+} extracellular medium (Fig. 1.10d, e). Moreover, at positive membrane potentials (+54 mV), adaptation was not observed after the photolysis of Nitr-5 (Kimitsuki and Ohmori 1992). This absence of adaptation at positive membrane potential might indicate that uncaging of Nitr-5 alone did not increase the Ca^{2+} concentration to the level sufficient to develop adaptation. Alternatively, these results suggested that the adaptation of MET current was not generated by the rise of intracellular basal Ca^{2+} concentration alone, but the Ca^{2+} entering through the MET channels somehow regulated the adaptation more efficiently than did the basal level of calcium ions in the cytosolic environment (Kimitsuki and Ohmori 1992).

The adaptation of MET current was reduced reversibly in a dihydrostreptomycin (DHSM, 20–50 μM) medium (Kimitsuki and Ohmori 1992). Interestingly, the change of adaptation induced by DHSM delayed to the change of MET current amplitude (Fig. 1.11). The amplitude of MET current changed preceding the change of adaptation, and after washout of DHSM, the speed and extent of adaptation were enhanced to the level more than that observed in the control before DHSM application (Fig. 1.11b). Moreover, time courses of disappearance of adaptation in low-Ca^{2+} (50–100 μM) medium and re-emergence of adaptation after the introduction of normal-Ca^{2+} (2.5 mM) medium, both lagged 10–20 s to the introduction of each extracellular medium (Kimitsuki and Ohmori 1992). The change of extracellular medium was completed within 3 s in the recording chamber when measured optically using phenol red in the medium. These delays suggest that Ca ions entering through the MET channel were required to be accumulated to reach certain concentration levels at the site of adaptation or they must pass through some buffering site

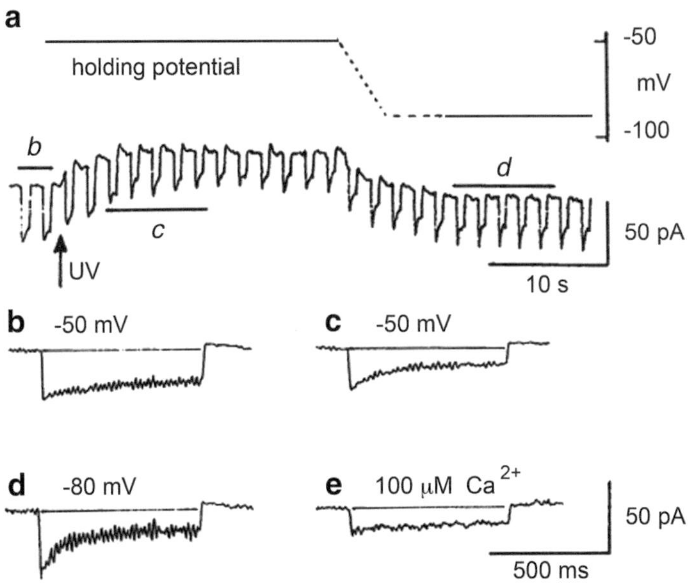

Fig. 1.10 Caged Ca^{2+} release affected adaptation of MET current
Caged Ca^{2+} compound Nitr5 was loaded into hair cell by using the membrane-permeable form (Nitr5-AM), and Ca^{2+} was released by a UV flash. (**a**) Hair cell was voltage clamped by using a KCl-based internal medium, and the MET current was induced repeatedly. UV flash on the Nitr-5-loaded hair cell at the timing of arrow induced a steady outward current. B, C, D, MET current before UV flash (**b**), after UV flash (**c**), and after negative shift of the holding potential to −80 mV in order to cancel the increase of holding current (**d**). Recording timings of **b**–**d** are indicated in the current trace of **a**. Note that adaptive decay of MET current was accelerated in C and D. Records in **a**–**d** were obtained in normal 2.5 mM Ca^{2+}-ACSF. (**e**) MET current recorded at −59 mV after **d** in low Ca^{2+} ACSF. Adaptive decay disappeared in **e**
From Kimitsuki and Ohmori (1992)

of Ca ions before reaching the adaptation site. There are many hypotheses proposed for the adaptation mechanism, but exact relationships are not certain yet between Ca ions that entered through the MET channel and the development of adaptation (Vollarth et al. 2007; Fettiplace and Kim 2014; Corns et al. 2014).

1.8 Membrane Excitability of the Hair Cell

Current through the MET channel generates receptor potential in the hair cell. The depolarizing receptor potential activates Ca^{2+} channels in the basolateral membrane and releases neurotransmitters, which activate auditory nerve fibers (ANFs). The level and speed of membrane depolarization is shaped by the activity of ion channels in hair cell and produces a unique response, such as the membrane potential

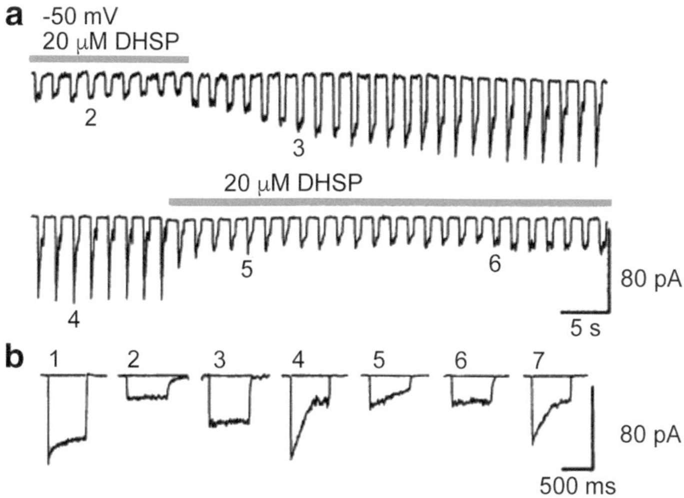

Fig. 1.11 DHSM affects MET current adaptation kinetics
(**a**) MET currents were induced at −50 mV repeatedly. The current amplitude was already reduced by preceding puff application of DHSM (20 μM). MET current numbered from 2 to 6 is extracted and illustrated in B. (**b**) Individual traces of MET current. DHSM blocked MET current (B2) to about 30% of the size recorded in the control (B1). After wash of DHSM, MET current amplitude was recovered (B3), and the adaptation kinetics gradually emerged with time (B3, B4). Second application of DHSM first reduced the size (B5), and then adaptation disappeared (B6). Thus, MET current size and the degree of adaptation are dissociated. B7 is after the wash of DHSM by the 2nd puff
From Kimitsuki and Ohmori (1992)

oscillation, which is interpreted as one form of the frequency tuning (Crawford and Fettiplace 1981a; Lewis and Hudspeth 1983).

Although chronologically, the following experiments about hair cell membrane excitability were conducted first in my laboratory before the recording of hair cell MET current, I preferred to follow the thread of signal transmission in this writing and described first the MET. The details of membrane excitability, modulation of excitability, and afferent synaptic transmission will follow in the following sections.

When I succeeded in isolation of hair cells from chick vestibular organ, I first studied ionic channels expressed in the hair cell membrane (Ohmori 1984b). Identification of hair cells among cell debris after the enzymatic treatment was easy because of the presence of hair bundle; a thornlike structure protruded from one pole of the cell body. I have chosen the vestibular organ for the first experiment, because the vestibular organ was easy to be found in the inner ear structure of the chick by the presence of otolith membrane. I thought hair cells must be there under the otolith. I peeled the otolith membrane, cut the sensory epithelium into small pieces, and isolated cells by enzymatic treatment and trituration. Isolated cells settled stably on a clean glass floor of a recording chamber.

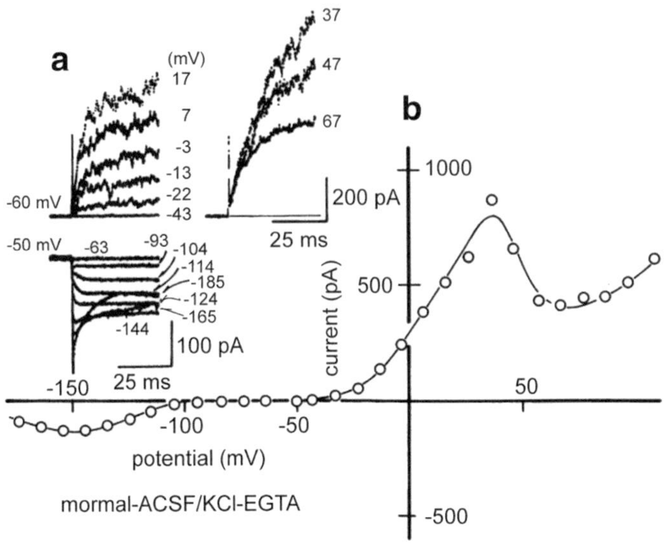

Fig. 1.12 Whole-cell currents from a hair cell
(**a**) Currents in normal ACSF bathing solution recorded with a KCl-EGTA-based internal medium. Current traces by step voltage changes to positive direction were recorded from the holding potential of −60 mV. Current amplitude was decreased at potentials more positive than +37 mV. Negative pulses were applied from the holding potential of −50 mV. Time-dependent current decay emerged at potentials more negative than −124 mV. (**b**) Steady-state amplitudes of the current are plotted against the membrane potential
From Ohmori (1984b)

In a standard ACSF (compositions in (mM) 155 NaCl, 5 KCl, 2.5 $CaCl_2$, 1 $MgCl_2$, 17 glucose, buffered to pH 7.4 by 10 Hepes-K), currents through anomalous rectifier K^+ channels, Ca^{2+}-activated K^+ channels, and Ca^{2+} channels were recorded, but no Na^+ channel currents were found when hair cells were whole-cell recorded using KCl-based or CsCl-based internal medium (155 KCl, 5 EGTA-K or 160 CsCl, 5 EGTA-Na). The I-V relation was N-shaped (Fig. 1.12b), indicating a presence of currents through Ca^{2+} channel and Ca^{2+}-activated K^+ channels.

1.8.1 Specific Capacitance of Hair Cell Membrane

The specific capacitance of hair cell was estimated as 1 μF/cm², a value consistent with a general consensus of the membrane capacitance (Ohmori 1984b). This estimation is based on the surface area and the input capacitance of the hair cell. The surface area of hair cell including the hair bundle was estimated as 738 μm², by summation of estimated surface area for the cell body and the hair bundle. A cylinder-shaped cell body of 9.5 μm in diameter and 13.6 μm in height contributed 486 μm². The hair bundle surface area was made up of stereocilia, of which height changed stepwise and the number and diameter were 90 and 0.25 μm (Takasaka and Smith 1971), respectively, and contributed to increase the surface area by 252 μm^2. The input capacitance measured was ranged from 4.5 to 7.4 pF, from the capacitive current in response to step changes of membrane potential. Since the capacitive current was likely underestimated because of limitation of the speed of recording amplifier, we took the largest capacitance as a typical value for the calculation of the specific capacitance of hair cell.

1.8.2 Ca^{2+} Channels

Property of hair cell calcium channel was basically the same as those of Ca^{2+} channels known at that time in the presynaptic axon of squid giant synapse (Llinas et al. 1981) and in neurosecretory cells (Fenwick et al. 1982; Hagiwara and Ohmori 1982). Ca^{2+}, Sr^{2+}, and Ba^{2+} were permeable through the channel. The channel was activated with a short delay after step voltage changes following the m^2 Hodgkin-Huxley kinetics and was not inactivated. Ba^{2+} ions were more permeable through the channel than other divalent cations. Ba^{2+} ions block K^+ channels and were very popular to be used in the study of Ca^{2+} channel kinetics. The steady-state noise of Ba^{2+} current demonstrated both low-frequency and high-frequency spectral components, which corresponded with the macroscopic gating kinetics and perhaps the fast open-close flickering kinetics of the channel, respectively. Based on the absolute size of Ca^{2+} current in single hair cells and the single Ca^{2+} channel current size estimated in other preparations, the number of Ca^{2+} channel was estimated as 200–500 per cell (Ohmori 1984b).

1.8.3 Anomalous Rectifier K^+ Channels

Anomalous rectifier K^+ channel in hair cells showed similar properties to those found in the oocyte membrane (Hagiwara and Takahashi 1974; Ohmori 1978) or in skeletal muscle (Gay and Stanfield 1977; Standen and Stanfield 1978, 1979). Single-channel conductance was dependent on the extracellular concentration of K^+ (23pS at 40 mM and 50 pS at 160 mM K^+ concentration) and followed nearly the square root of extracellular K^+ concentration (Ohmori 1984b). This is the property found in anomalous rectifier K^+ channel in other cells (Hagiwara and Takahashi 1974; Ciani et al. 1978).

1.8.4 Ca^{2+}-Activated K^+ Channels

There are tall and short hair cells in the chicken cochlea (Fig. 1.13). The tall hair cell is homologous to the mammalian inner hair cell and receives a rich innervation of afferent fibers, while the short hair cell receives limited number of afferent

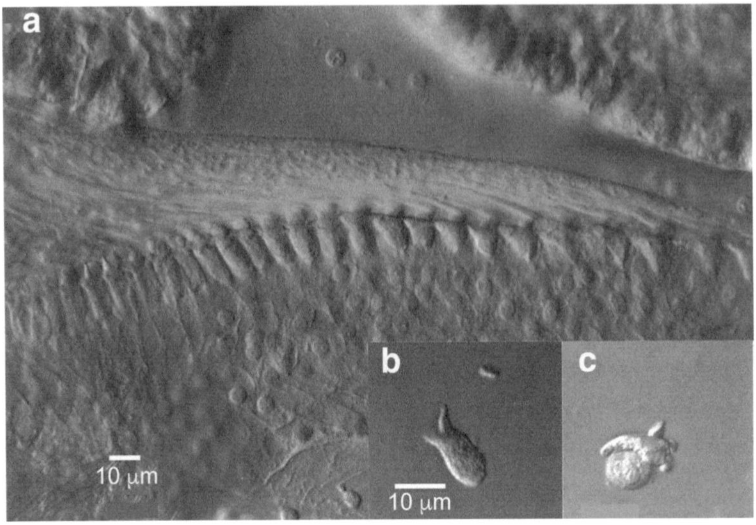

Fig. 1.13 Transverse section of the basilar papilla of chick cochlea organ
(**a**) Hair cells stand in line under the tectorial membrane. Note that fibers extending from the tectorial membrane made contact with the tip of hair bundle. Tall hair cells are located toward the left (the proximal part of the basilar membrane). Short hair cells are in the distal location (toward right) and have well-developed cuticular plate. (**b**) Tall hair cell. (**c**) Short hair cell
From Shigemoto and Ohmori (1991)

innervations but a large number of efferent innervations. Accordingly, the short hair cell is homologous to the outer hair cell of the mammalian cochlea. As it is suggested by the N-shaped current-voltage relationship (Fig. 1.12b), hair cells have the Ca^{2+}-activated K^+ channel (Ohmori 1984b). Tan et al. (2013) found electrical resonance property of short hair cells through the activity of Ca^{2+}-activated K^+ channels. Electrical tuning property of hair cell has been discussed in the turtle (Crawford and Fettiplace 1981a) and in the frog sacculus (Lewis and Hudspeth, 1983; Ashmore 1983). It is discussed that Ca^{2+} channels, Ca^{2+}-activated K^+ channels, and, in some case, A-type K^+ channels are responsible for the electrical resonance of hair cell membrane. The electrical tuning seems important in the frequency range lower than 1 kHz because of the limitation of hair cell membrane time constant, which, however, seems to have physiological importance in ground-crawling animals.

1.8.5 *Distribution of Ion Channels*

Although I could not detect Na^+ channel current from hair cells of the chick (Ohmori 1984b), Na^+ channel currents were recently found in hair cells of mammalian species (Marcotti et al. 2003; Eckrich et al. 2012) and in immature hair cells of cultured rat organ of Corti (Oliver et al. 1997). Na^+ currents were found even in the chicken semicircular canal hair cells (Masetto et al. 2003).

MET channels are so far the only channel distributed in the apical pole of the hair cell. Ca^{2+} channels, Ca^{2+}-activated K^+ channels, and anomalous rectifier K^+ channels are likely distributed in the basolateral membrane. Thus, the distribution of ion channels is uneven on the hair cell membrane. The apical surface of hair cell is exposed to the endolymph that has rich K^+ composition, while the basolateral surface is bathed in the perilymph that has ionic compositions similar to the normal extracellular fluid. Accordingly, the resting membrane potential of −62 ± 7 mV was created and maintained across the basolateral membrane between the intracellular medium and perilymph (Ohmori 1984b), while the external surface of MET channel faced the endolymph and the positive endolymphatic potential (up to 100 mV; Tasaki et al. 1954; Bosher and Warren 1971; Steel and Barkway 1989; Gill and Salt 1997). The driving force for MET current was more than doubled by the presence of endolymphatic potential.

1.9 Modulation of Hair Cell Membrane Excitability

1.9.1 Cholinergic Hyperpolarization of Hair Cell Membrane

There are two distinct hair cell types in the mammalian cochlear: the inner and outer hair cells. Innervation of efferent fibers dominated on outer hair cells but poor on inner hair cells (Spoendlin 1972; Webster 1992). Olivocochlear bundle (OCB) is the major source of efferent innervation (Spangler and Warr 1991, Warr 1992; Fujno et al. 1997). It is well known that the electrical stimulation applied to the OCB at the floor of fourth ventricle induced suppression of sound-evoked auditory nerve fiber discharges (Galambos 1956; Furukawa 1981).

Inhibition of auditory-evoked postsynaptic response was reported by efferent electrical stimulation in hair cells of vestibular organs and lateral line organs in several species (Flock and Russell 1976; Furukawa 1981; Ashmore and Russell 1983). Effects of electrical stimulation applied to the efferent fibers were studied in details in the turtle hair cells. A transient hyperpolarization response was generated, and the response reversed polarity at around −80 mV, at the K^+ reversal potential (Art et al. 1984). Single shock of the efferent fiber generated a small hyperpolarization, and the hyperpolarization was accumulated when a train of stimulation was applied. Application of acetylcholine induced a transient membrane hyperpolarization, while curare or atropine abolished the postsynaptic response. More interestingly, the postsynaptic potential was accompanied with an early depolarizing synaptic component. Early current was also reported later in the short hair cells of the chicken by a fast puff application of ACh (Fuchs and Murrow 1992).

This early synaptic current was attributed to the activation of α-9 ACh receptor (Elgoyhen et al. 1994). Alpha-9 ACh receptors were found by cloning technology from rat genomic library. Alpha-9 ACh receptors were ionotropic and cation permeable, including Ca^{2+}. Transient increase of intracellular Ca^{2+} concentration activated the Ca^{2+}-activated K^+ conductance and hyperpolarized the hair cell membrane. The α-9 ACh receptor had mixed natures of nicotinic and muscarinic receptors, because it was blocked by d-tubocurarine (blocker of nicotinic ACh receptor), by α-bungarotoxin (nicotinic antagonist), by κ-bungarotoxin (nicotinic antagonist), by strychnine (antagonist of inhibitory synapse, GABA or glycine receptor), and moreover by atropine (blocker of muscarinic ACh receptor). Both inner hair cells and outer hair cells have α-9 ACh receptors.

1.9.2 Slow Membrane Hyperpolarization

Different from the mammalian cochlea, avian cochlea has many hair cells in a transverse section; about 20–50 hair cells are found in one section (Fig. 1.13a). A large number of hair cells were located near the lagenar region (Takasaka and Smith 1971). Hair cells gradually changed shape toward the periphery of basilar membrane. Tall hair cells are located closer to the proximal part of the basilar membrane and short hair cells in the distal part (Fig. 1.13a) (Takasaka and Smith 1971; Tanaka and Smith 1978). Short hair cells receive rich efferent innervations. In contrast the efferent innervation to tall hair cells is trivial.

In short hair cells, ACh response was robust both in calcium signal and in membrane current, in contrast to the small signals found in tall hair cells (Shigemoto and Ohmori 1991). ACh puff application (100 μM) induced a prolonged membrane hyperpolarization (Fig. 1.14). ACh-induced current was reversed at −85 ± 4.2 mV, and the reversal potential was dependent on the extracellular K^+ concentration. ACh-induced K^+ current was blocked by atropine (1 μM, blocker of muscarinic receptor) or quinine (100 μM, blocker of Ca^{2+}-activated K^+ channel) and can be still

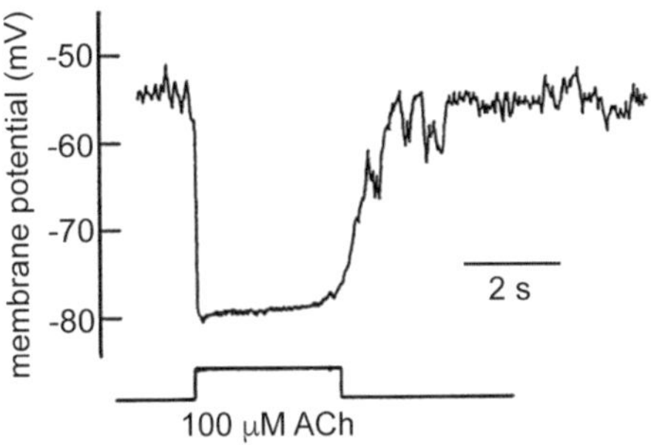

Fig. 1.14 ACh-induced membrane hyperpolarization
Puff application of ACh (100 μM) induced hyperpolarization of hair cell membrane from −54 mV to −80 mV. A short hair cell was current clamped by using a KCl-EGTA (0.1 mM)-based internal medium
From Shigemoto and Ohmori (1991)

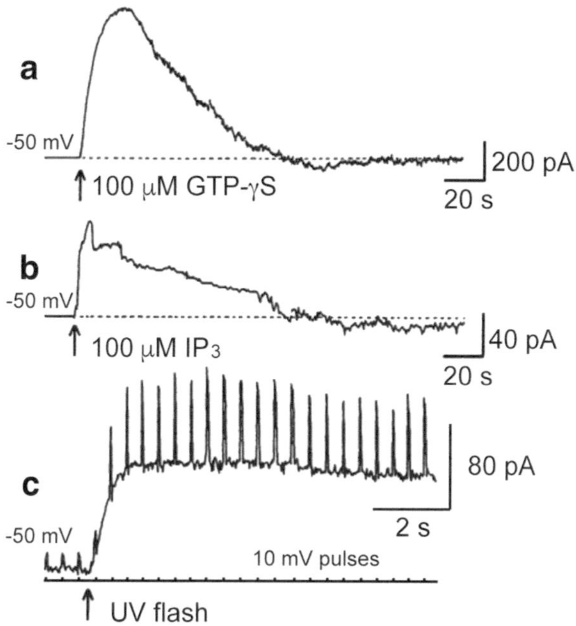

Fig. 1.15 GTPγS, IP_3, and intracellular Ca^{2+} induced outward current in short hair cells (**a**, **b**) Patch electrode was cell attached on a short hair cell membrane first in this recording, and the whole-cell recording was started by the rupture of patch membrane at the timing of arrow. The patch pipette contained GTPγS (100 μM) in **a** and IP_3 (100 μM) in **b**. (**c**) After incubation with a membrane-permeable form of caged Ca^{2+} compound (Nir5-AM), hair cell was whole cell clamped by using a KCl-EGTA (0.1 mM)-based internal medium. UV flash released Ca^{2+} intracellularly at the timing of arrow. 10 mV depolarizing pulses were applied repeatedly from −50 mV in order to monitor the membrane conductance. After the UV flash, membrane conductance was increased From Shigemoto and Ohmori (1991)

generated in a Ca^{2+}-free extracellular medium, although the amplitude was reduced nearly to the half of control response in the normal ACSF.

Intracellular injection of GTPγS, IP_3, or photoactivation of a caged Ca^{2+} compound (Nitr-5) induced outward current at −50 mV (Fig. 1.15), while high concentration of intracellular EGTA reduced the ACh-induced current responses. Accordingly, ACh induced the K^+ conductance by releasing Ca^{2+} intracellularly through the activity of G protein-coupled signaling molecules in a sequence, such as PLC (phospholipase C) and IP_3 (Shigemoto and Ohmori 1991). K^+ current was activated with a short delay (1 ms after ionophoresis of ACh), which could be due partly to the Ca^{2+} influx via α-9 ACh receptors (Fuchs and Murrow 1992; Elgoyhen et al. 1994).

Under the cell-attached patch recording on short hair cells, single-channel current activity was recorded in high percentage (58%) when ACh (100 μM) was included within the pipette solution, but the channel activity occurred in low percentage (4.5%) when ACh was applied to the nearby membrane (Shigemoto and

Ohmori 1991). This may indicate that ACh receptors and Ca^{2+}-activated K^+ channels are localized close to each other in the hair cell membrane. ACh-induced calcium responses and membrane hyperpolarization were observed in the isolated outer hair cells of the guinea pig (Doi and Ohmori 1993) and in bullfrog saccular hair cells (Yoshida et al. 1994). In contrast in avian tall hair cells, a small and very slow outward current was generated with a long delay when ACh was puff applied.

Accordingly, cholinergic innervation on hair cells had two phenotypes: (1) Early effects through the activation of α-9 ACh receptor. Ca^{2+} influx through α-9 ACh receptors activated the Ca^{2+}-activated K^+ channel and hyperpolarized the hair cell membrane transiently in response to efferent stimulation. (2) The second phenotype was more robust and prolonged the membrane hyperpolarization. The cholinergic muscarinic activation released Ca^{2+} intracellularly and generated the slow and prolonged membrane hyperpolarization, which induced a long-lasting inhibition of afferent synaptic transmission in response to the efferent stimulation. The slow ACh-induced calcium response observed in these hair cells was somewhat consistent with the report of slow cholinergic effects on hair cells by stimulation of the olivocochlear bundle in guinea pig, which was sensitive to cholinergic antagonists (Sridhar et al. 1995).

It should be noted here that activation of K^+ channels, in general, has inhibitory effects on the receptor potential. One clear example is the K^+ channel encoded by KCNQ4 gene, which is expressed in hair cells and is associated with deafness, the progressive dominant hearing loss. The channel is activated near the resting potential and attenuates the receptor potential (Holt et al. 2007).

1.9.3 Purinergic Sensitivity of Hair Cell Membrane

ATP also induced Ca^{2+} response in hair cells, and the response was in two phases, fast and slow kinetics (Shigemoto and Ohmori 1990). The fast phase accompanied with inward current and disappeared in Ca^{2+}-free medium and likely reflected the influx of Ca^{2+} from extracellular medium (Shigemoto and Ohmori 1990). Cationic currents induced by ATP application were studied in detail in outer hair cells of the guinea pig (Nakagawa et al. 1990). Various cations were permeable through the channel (Li (P_X/P_{Cs} = 1.23) > Na (1.08) > Cs (1.0) >> Choline (0.07)), and the highest permeability was of Ca ions (3.02). The relative potency of agonistic effect was the highest for ATP followed by other ATP analogues (ADP, α, β-meATP, α, γ-meATP). ATP application generated inward current and slow Ca^{2+} responses even in a Ca^{2+}-free extracellular medium in outer hair cells of the guinea pig (Ashmore and Ohmori 1990). Supporting cells of the organ of Corti were also sensitive to ATP and increased the intracellular calcium level. In these cells, rise of intracellular Ca^{2+} occurred even in the absence of extracellular Ca^{2+}. Accordingly, the organ of Corti had receptors for ATP on various cell types. ATP opened cation-permeable channels and induced entry of Ca^{2+} and also mobilized Ca^{2+} from intracellular stores.

Purinergic P2X2 receptor channels were crucial in adaptation of hair cells to intense sound stimulus (Houseley et al. 2013).

Because purinergic receptor channels P2X-Rs transport cations including calcium ions, it is possible that their activity influences neural excitability and the calcium-dependent metabolic processes. In developing auditory circuits of Mongolian gerbil, ATP was released in the cochlea, activated hair cells, and triggered bursts of action potentials in the auditory nerve fibers (Dietz et al. 2012). ATP increased both spontaneous and acoustically evoked firing activity in neurons of cochlear nucleus and the firing activity accompanied with the increase of intracellular Ca^{2+}. These ATP effects occurred from prehearing to early stages of hearing but diminished after maturity when ATP release was suspended.

Purinergic modulation of neural activity should be important, on the spontaneous firing activity and Ca^{2+} responses. Ca^{2+} responses activated protein kinase C and downstream metabolic processes and trigger long-term plasticity. These ATP effects could possibly play essential roles in establishment of functional auditory circuits during the critical period of development, via providing signals for survival of target neurons, for tonotopic refinement of afferent connections, and for adjustment of synaptic strength (Rubel and Fritzsh 2002; Leake et al. 2006; Kandler et al. 2009; McKay and Oleskevich 2007).

1.9.4 Modulation of Hair Cell Excitability by Other Agents

Hair cells incubated with CGRP (1 μM, calcitonin gene-related peptide) increased the sensitivity to puff-applied ACh that induced Ca^{2+} influx in the chicken hair cells (Shigemoto and Ohmori 1990). Application of CGRP itself did not induce Ca^{2+} response. CGRP was immunolocalized in the efferent terminals in the lateral line organ of *Xenopus laevis* (Adams et al. 1987). Physiological roles of CGRP were not certain, but in the lateral line organ, application of CGRP increased the firing rate of afferent fibers (Sewell and Starr 1991). CGRP coexisted with ACh in motor nerve terminals and acted as a trophic factor to increase the synthesis of ACh receptors in motor neurons (New and Mudge 1986). CGRP co-released with ACh from efferent terminals might facilitate firing activity of auditory afferent fibers. Presence of CGRP was demonstrated at the synapse region of the outer hair cells and efferent fibers of olivocochlear bundle of rats (Kitajiri et al. 1985; Takeda et al. 1986).

1.10 Hair Cell Afferent Synaptic Transmission

Depolarization of hair cell induces action potential in the postsynaptic afferent nerve fibers (Furukawa and Ishii 1967). Many trials have been carried out to identify the neurotransmitter released from hair cells, and several candidates were proposed before we conducted an experiment to detect release of glutamate from hair cells.

GABA (Flock and Lam 1974), glutamate (Bobbin 1979, Comis and Leng 1979), catecholamines (Thornhill 1972), and even unidentified neuroactive substances (Sewell et al. 1978) were proposed as afferent neurotransmitter. However, there was no direct evidence of release of these substances in those preceding experiments.

Since we knew that cultured cochlear ganglion cells have receptors for glutamate (Yamaguchi and Ohmori 1990), we thought that glutamate was the most likely candidate of neurotransmitter released from hair cells. We planned experiments to identify the release of glutamate from hair cells (Kataoka and Ohmori 1994). We had primary culture of rat cerebellum in the laboratory. We thought to use these cultured neurons as detectors for released substances by hair cell depolarization. Purkinje cell was first used as a detector. Purkinje cell was large and was easy to be whole-cell patch clamped (Fig. 1.16). Unfortunately, no current response occurred in that neuron on depolarization of closely apposed hair cell that was also whole-cell voltage clamped. We thought transmitter release machinery might have been disrupted by perfusion within the cell as a consequence of whole-cell recording. Thus, we used nystatin to voltage clamp these cells to prevent possible dilution of the cell interior. One day we realized by reading a review article on glutamate receptors that glutamate sensitivity was different between AMPA-type receptors and NMDA-type receptors (Mayer and Westbrook 1987). NMDA receptors had higher affinity to glutamate. Moreover, matured Purkinje cells were unique in not having the NMDA receptors but having the non-NMDA receptors (Llano et al. 1991; Rosenmund et al. 1992). In contrast, cerebellar granule cells had rich expression of NMDA receptors.

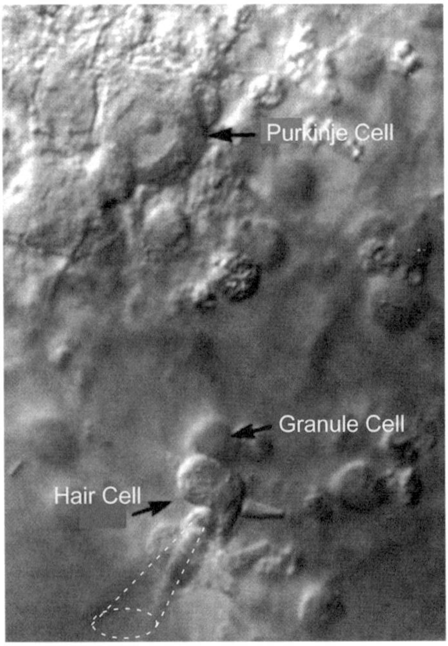

Fig. 1.16 Photomicrograph of primary cultured rat cerebellum and hair cell Isolated hair cell was transported by holding at the cuticular plate by a pipette (broken lines) and was abutted to a granule cell. Hair cells were isolated from the chick, and Purkinje cells and granule cells were primary cultured from rat pups From Kataoka and Ohmori (1994)

We therefore tried to use granule cells as detectors of released transmitters (Kataoka and Ohmori 1994).

An isolated hair cell from chicken cochlea was gently sucked and transferred by a holder pipette and was placed in close contact with a cultured granule cell of the rat cerebellum (Fig. 1.16). Both cells were whole-cell voltage clamped by utilizing a nystatin-perforated patch recording technique (Fig. 1.17). Upon depolarization of the hair cell (HC), current was induced in the granule cell (GrC) (Fig. 1.17a). The granule cell was voltage clamped at +55 mV to eliminate the Mg^{2+} block of NMDA receptor channel (Nowak et al. 1984) and simultaneously to increase the driving force of current through the channels. Induced current in granule cells was reversibly suppressed by a local application of 2-amino-5-phosphonovalerate (APV, Fig. 1.17b, c), which indicated that the current was generated through activation of the NMDA subtype of glutamate receptors (Kataoka and Ohmori 1994). Current amplitude of the granule cell was dependent on the level of hair cell depolarization (Fig. 1.18) and was progressively increased from –20 to +10 mV step depolarization of the hair cell. At more positive potentials, the current amplitude was decreased. This voltage dependence was similar to the Ca^{2+} currents of hair cells (Ohmori 1984b). Furthermore, when Ca^{2+} concentration was increased in the hair cell by UV irradiation of a caged Ca^{2+} compound (Nitr-5), the closely apposed granule cell-generated outward current when recorded at +55 mV. These observations indicated that the hair cell released the neurotransmitter when intracellular Ca^{2+} concentration was raised upon depolarization and that the likely neurotransmitter was glutamate (Kataoka and Ohmori 1994).

It has been reported that several endogenous excitatory amino acids (EAAs, L-glutamate, L-aspartate, and L-homocysteate) activate the NMDA subtype of glutamate receptors (Curras and Dingledine 1992; Kilic et al. 1992). We therefore conducted experiments to show which was the most likely neurotransmitter released from hair cells, by utilizing the sensitivity difference between granule cells and Purkinje cells to EAAs (Kataoka and Ohmori 1996). Granule cells had rich expression of NMDA glutamate receptors, while Purkinje cells had non-NMDA-type glutamate receptors (Llano et al. 1991; Rosenmund et al. 1992). We measured concentration-response relationships for these three endogenous EAAs in the granule cell and in the Purkinje cell. The granule cell was most sensitive to glutamate, whereas these three EAAs did not show any sensitivity difference in the Purkinje cell. From the dose-response relationships, the half maximum responses in Purkinje cells were at 17–60 μM for glutamate, aspartate, and homocysteate when EAAs were puff applied. In contrast, the half maximum response was found at the concentration of 1.1 μM for glutamate, 11 μM for aspartate, and 10 μM for homocysteate in the granule cell. In the granule cell voltage clamped at +55 mV, depolarization of closely abutted hair cell induced in average 70 pA outward current (70 ± 33 pA, $n = 8$). This current level was equivalent to the size of current induced by puff application of 0.33 μM glutamate, 5 μM aspartate, or 2 μM homocysteate when measured from the dose-response relationships in granule cells. However, this critical concentration (0.33 μM) of glutamate failed to induce any current in Purkinje cells voltage clamped at -60 mV. Nevertheless, the critical concentration of other two

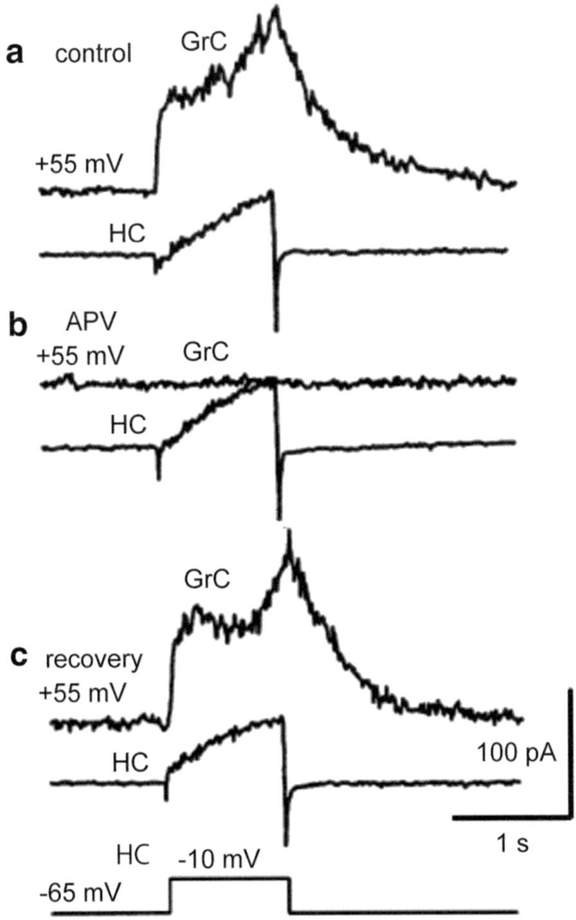

Fig. 1.17 Glutamate released from hair cell activated current in granule cell
(**a**) A large outward current was induced in the granule cell voltage clamped at +55 mV by the abutted hair cell membrane depolarization to −10 mV from the −65 mV holding potential. Protocol of voltage steps of the hair cell is indicated at the bottom of **c**. Hair cell depolarization induced a small inward current and a larger outward current in the hair cell. Granule cell was voltage clamped at +55 mV to release the Mg^{2+} block of NMDA receptor channel. (**b**) Local application of APV (200 μM) by a puff suppressed currents in the granule cell. (**c**) Granule cell current was restored after rinsing APV with 10 mM Ca^{2+} saline for 40 s
GrC granule cell, *HC* hair cell
From Kataoka Ohmori (1994)

amino acids (5 μM aspartate or 2 μM homocysteate) evoked inward currents in the Purkinje cell. Accordingly, we concluded that glutamate is the most likely neurotransmitter released from hair cells of the chick (Kataoka and Ohmori 1996).

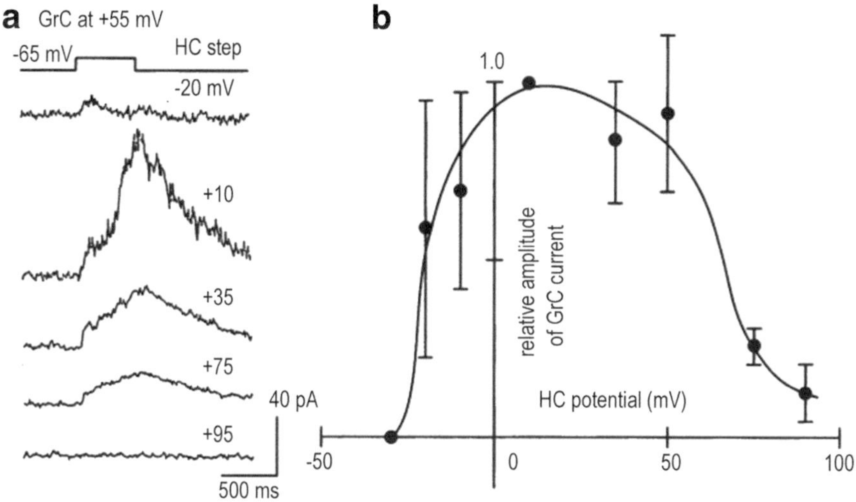

Fig. 1.18 Dependence of granule cell current on the hair cell membrane depolarization (**a**) Granule cell was voltage clamped at +55 mV, and the membrane potential of abutted hair cell was step changed from −65 mV holding potential to the potentials indicated. The granule cell current was increased with the increase of hair cell step depolarization to +10 mV and then decreased when the hair cell depolarization was larger. (**b**) Granule cell current amplitude was normalized and was plotted against the hair cell membrane potential
From Kataoka and Ohmori (1994)

Release of glutamate was later reported in the mouse hair cell by recording from the bouton of afferent synapse terminal that still remained after the hair cell isolation procedure (Glowatzki and Fuchs 2002). Spontaneous synaptic currents were recorded from the synapse bouton, and the activity was blocked by CNQX (antagonist of glutamate receptors), reversibly.

1.11 Spiral Ganglion Neurons

Using cultured spiral ganglion neurons from the chick embryo (E16-19), we investigated ionic currents (Yamaguchi and Ohmori 1990). Cochlear ganglion (CG) cells had a shape of pseudo-bipolar neuron after culture for 1–14 days (Fig. 1.19). Currents through TTX-sensitive Na^+ channel, two types of Ca^{2+} channels (low-voltage-activated Ca^{2+} channels and high-voltage-activated non-inactivating Ca^{2+} channels), and NMDA subtype of glutamate receptors were found. Pseudo-bipolar shape of the neuron seems advantageous in fast speed conduction of spikes from the peripheral to central neurite (the axon), by evading the capacitive load induced by somatic excitation (Hossain et al. 2005).

The glutamate response was depressed by 1 mM Mg^{2+} in a voltage-dependent manner at negative membrane potentials, and the majority (84%) of glutamate

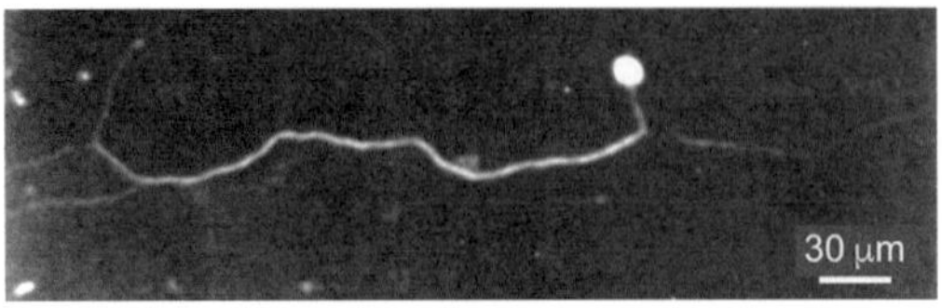

Fig. 1.19 Primary cultured cochlear ganglion neuron from embryonic chicken
Pseudo-bipolar neuron was visualized by immunostaining against neurofilament. Calibration bar 10 μm
From Yamaguchi and Ohmori (1990)

response was eliminated by 0.1 mM APV (Yamaguchi and Ohmori 1990). Thus, NMDA receptors were the major glutamate receptor components of the synapse between hair cells and auditory afferent fibers in late embryonic age of the animal. However, the slow gating kinetics of NMDA receptors may not be appropriate for this acoustic synapse which needs to operate at fast speed. It is reported that NMDA receptors were largely replaced with AMPA-type receptors during development in rat bushy cell of AVCN (anteroventral cochlear nucleus; Bellingham et al. 1998; Beurg et al. 2016). There is also a general consensus that the expression level of NMDA receptors is lower in adult animals than in animals of early developmental days (McBain and Mayer 1994). Accordingly, the expression of NMDA receptors in the cultured CG neurons may reflect an early developmental phase of the synapse between hair cell and auditory afferent nerve fibers.

We also found cholinergic inhibitory effects on K^+ channels by carbamylcholine (CCh) in the cultured CG neuron (Yamaguchi and Ohmori 1993). Atropine (3 μM) and pirenzepine (3 μM) blocked the suppressive effect of CCh on K^+ current, suggesting that the K^+ current was modulated by M1 muscarinic receptors. This suppression of K^+ current was enhanced by GTP-γS (0.1 mM), suggesting some involvement of GTP-binding protein in the cholinergic modulation. Moreover, the suppression of K^+ current by CCh was mimicked by activators of protein kinase C, such as OAG (1-oleoyl-2-acetyl-sn-glycerol), PDBu (phorbol dibutyrate), and PMA (phorbol 12-myristate 13-acetate). The protein kinase inhibitor, staurosporine, applied internally through the recording patch pipette reduced the suppression of K^+ current by CCh (Yamaguchi and Ohmori 1993). This block of K^+ channel by muscarinic receptors may contribute to increase the input impedance of afferent fibers, which seems consistent with the facilitating effect of efferent stimulation on auditory nerve transmission (Goldberg and Fernandez 1980). In the lateral line organ of *Xenopus laevis*, efferent stimulation induced a rapid increase of afferent discharges that were followed by inhibition and slow increase of afferent discharges (Sewell and Starr 1991). These rapid changes of afferent fiber discharges were likely the effect of cholinergic synapse, and the slow increase of firing activity might be related to the facilitating effects of co-released CGRP from efferent terminals (Shigemoto and Ohmori 1990).

In contrast to the observation of Na^+ current in the soma of cultured CG cell of the chick embryo, immunohistological expression level of Na^+ channel was trivial in the bipolar-shaped CG cell soma of the matured mouse (Hossain et al. 2005). There are two types of CG neurons in the mouse. The type I CG cell innervates single inner hair cell, and the type II neuron innervates multiple outer hair cells. Na^+ channel was not detected by immunostaining (panNav) in the cell body of either types. Instead, Nav1.6 channels were detected at the peripheral and central hemi-nodes that flanked the cell body in the type I neuron. At the nodes of Ranvier and in peripheral and central axon initial segments of type II neurons, Nav1.6 channels were detected. Accordingly, it was expected that across the two spike initiation sites on the axon that flanked the cell body of CG neuron, action potential should be transmitted from peripheral to central axons without inducing spikes in the cell body. This mode of Na^+ channel expression is therefore important to overcome the large capacitive load of cell body and to ensure the fast conduction speed of action potentials from hair cell to the brainstem nucleus (Hossain et al. 2005). Nav1.6 immunostainings were also detected in terminal regions of the axon that make contact with hair cells in both types of CG neurons, which likely facilitated transmission of spikes from hair cell afferent synapse along the axon to CG neurons.

1.12 Recapitulation of this Chapter

Airborne signal is transduced to electrical signal in hair cells. Hair bundle displacement gates MET channels and generates electrical signal, which makes it possible for the nervous system to handle the auditory information. The first electrical signal is the depolarizing receptor potential in hair cells, which activates Ca^{2+} channels in the basolateral membrane to elevate the Ca^{2+} concentration in hair cells. Glutamate, the afferent neurotransmitter, is released from the hair cell and triggers spikes in the axon of cochlear ganglion (CG) cells. CG cells issue the auditory nerve fibers (ANFs). Spikes in ANFs are conducted along the axon, but the spike bypasses the cell body. A large capacitive load of CG cell soma is a trouble for a fast spike conduction. Conveniently, the cell soma was practically made unexcitable by the absence of Na^+ channels. Hair cell receptor potential is a target of efferent modulation. By efferent signals the cholinergic receptors on hair cells are activated to increase intracellular Ca ions, which activate the Ca^{2+}-activated K^+ channels to generate a prolonged membrane hyperpolarization to suppress the afferent synapse transmission. In contrast, the cholinergic effects on ANFs are facilitatory by reducing K^+ channel activity. Thus, the hair cell sensitivity to mechanical stimulation and the transmission to afferent synapse and moreover the spike generation in ANFs could be under the control of higher-order auditory nuclei.

References

Adams JC, Mroz EA, Sewell WF (1987) A possible neurotransmitter role for CGRP in a hair-cell sensory organ. Brain Res 419:347–351

Art JJ, Fettiplace R, Fuchs PA (1984) Synaptic hyperpolarization and inhibition of turtle cochlear hair cells. J Physiol 356:525–550

Ashmore JF (1983) Frequency tuning in a frog vestibular organ. Nature 304:536–538

Ashmore JF, Ohmori H (1990) Control of intracellular calcium by ATP in isolated outer hair cells of the guinea-pig cochlea. J Physiol 428:109–131

Ashmore JF, Russell IJ (1983) Sensory and effector functions of vertebrate hair cells. J Submicrosc Cytol 15(1):163–166

Bagger-Sjoebaeck D, Wersaell J (1973) The sensory hairs and tectorial membrane of the basilar papilla in the lizard Calotes versicolor. J Neurocytol 2:329–350

Bellingham MC, Lim R, Walmsley B (1998) Developmental changes in EPSC quantal size and quantal content at a central glutamatergic synapse in rat. J Physiol 511:861–869

Beurg M, Evans MG, Hackney CM, Fettiplace R (2006) A large-conductance calcium-selective mechanotransducer channel in mammalian cochlear hair cells. J Neurosci 26(43):10992–11000

Beurg M, Fettiplace R, Nam JH, Ricci AJ (2009) Localization of inner hair cell mechanotransducer channels using high-speed calcium imaging. Nat Neurosci 12(5):553–558

Beurg M, Kim KX, Fettiplace R (2014) Conductance and block of hair-cell mechanotransducer channels in transmembrane channel-like protein mutants. J Gen Physiol 144(1):55–69

Beurg M, Goldring AC, Ricci AJ, Fettiplace R (2016) Development and localization of reverse-polarity mechanotransducer channels in cochlear hair cells. Proc Natl Acad Sci U S A 113(24):6767–6772

Bobbin RP (1979) Glutamate and aspartate mimic the afferent transmitter in the cochlea. Exp Brain Res 34(2):389–393

Bosher SK, Warren RL (1971) A study of the electrochemistry and osmotic relationships of the cochlear fluids in the neonatal rat at the time of the development of the endocochlear potential. J Physiol 212(3):739–761

Bosher SK, Warren RL (1978) Very low calcium content of cochlear endolymph, an extracellular fluid. Nature 273(5661):377–378

Ciani S, Krasne S, Miyazaki S, Hagiwara S (1978) A model for anomalous rectification: electrochemical-potential-dependent gating of membrane channels. J Membr Biol 44(2):103–134

Comis SD, Leng G (1979) Action of putative neurotransmitters in the guinea pig cochlea. Exp Brain Res 36(1):119–128

Corey DP, Holt JR (2016) Are TMCs the Mechanotransduction channels of vertebrate hair cells? J Neurosci 36(43):10921–10926

Corey DP, Hudspeth AJ (1979) Ionic basis of the receptor potential in a vertebrate hair cell. Nature 281(5733):675–677

Corey DP, Hudspeth AJ (1983) Analysis of the microphonic potential of the bullfrog's sacculus. J Neurosci 3(5):942–961

Corns LF, Johnson SL, Kros CJ, Marcotti W (2014) Calcium entry into stereocilia drives adaptation of the mechanoelectrical transducer current of mammalian cochlear hair cells. Proc Natl Acad Sci U S A 111(41):14918–14923

Crawford AC, Fettiplace R (1981a) An electrical tuning mechanism in turtle cochlear hair cells. J Physiol 312:377–412

Crawford AC, Fettiplace R (1981b) Non-linearities in the responses of turtle hair cells. J Physiol 315:317–338

Crawford AC, Evans MG, Fettiplace R (1989) Activation and adaptation of transducer currents in turtle hair cells. J Physiol 419:405–434

Crawford AC, Evans MG, Fettiplace R (1991) The actions of calcium on the mechano-electrical transducer current of turtle hair cells. J Physiol 434:369–398

Curras MC, Dingledine R (1992) Selectivity of amino acid transmitters acting at N-methyl-D-aspartate and amino-3-hydroxy-5-methyl-4-isoxazolepropionate receptors. Mol Pharmacol 41(3):520–526

Denk W, Strickler JH, Webb WW (1990) Two-photon laser scanning fluorescence microscopy. Science 248(4951):73–76

Denk W, Holt JR, Shepherd GM, Corey DP (1995) Calcium imaging of single stereocilia in hair cells: localization of transduction channels at both ends of tip links. Neuron 15(6):1311–1321

Dietz B, Jovanovic S, Wielsch B, Nerlich J, Rübsamen R, Milenkovic I (2012) Purinergic modulation of neuronal activity in developing auditory brainstem. J Neurosci 32(31):10699–10712

Doi T, Ohmori H (1993) Acetylcholine increases intracellular Ca2+ concentration and hyperpolarizes the guinea-pig outer hair cell. Hear Res 67(1–2):179–188

Eatock RA, Corey DP, Hudspeth AJ (1987) Adaptation of mechanoelectrical transduction in hair cells of the bullfrog's sacculus. J Neurosci 7(9):2821–2836

Eckrich T, Varakina K, Johnson SL, Franz C, Singer W, Kuhn S, Knipper M, Holley MC, Marcotti W (2012) Development and function of the voltage-gated sodium current in immature mammalian cochlear inner hair cells. PLoS One 7(9):e45732

Elgoyhen AB, Johnson DS, Boulter J, Vetter DE, Heinemann S (1994) Alpha 9: an acetylcholine receptor with novel pharmacological properties expressed in rat cochlear hair cells. Cell 79(4):705–715

Eyring H (1935) The activated complex in chemical reactions. J Chem Phys 3:107–115

Fenwick EM, Marty A, Neher E (1982) A patch-clamp study of bovine chromaffin cells and of their sensitivity to acetylcholine. J Physiol 331:577–597

Fettiplace R, Kim KX (2014) The physiology of mechanoelectrical transduction channels in hearing. Physiol Rev 94(3):951–986

Flock A (1965) Transducing mechanisms in the lateral line canal organ receptors. Cold Spring Harb Symp Quant Biol 30:133–145

Flock A, Lam DMK (1974) Neurotransmitter synthesis in inner ear and lateral line sense organs. Nature 249:142–144

Flock A, Russell I (1976) Inhibition by efferent nerve fibres: action on hair cells and afferent synaptic transmission in the lateral line canal organ of the burbot Lota lota. J Physiol 257(1):45–62

Fuchs PA, Murrow BW (1992) Cholinergic inhibition of short (outer) hair cells of the chick's cochlea. J Neurosci 12(3):800–809

Fujino K, Koyano K, Ohmori H (1997) Lateral and medial olivocochlear neurons have distinct electrophysiological properties in the rat brain slice. J Neurophysiol 77(5):2788–2804

Furness DN, Hackney CM, Benos DJ (1996) The binding site on cochlear stereocilia for antisera raised against renal Na+ channels is blocked by amiloride and dihydrostreptomycin. Hear Res 93(1–2):136–146

Furness DN, Zetes DE, Hackney CM, Steele CR (1997) Kinematic analysis of shear displacement as a means for operating mechanotransduction channels in the contact region between adjacent stereocilia of mammalian cochlear hair cells. Proc Biol Sci 264(1378):45–51

Furukawa T (1981) Effects of efferent stimulation on the saccule of goldfish. J Physiol 315:203–215

Furukawa T, Ishii Y (1967) Neurophysiological studies on hearing in goldfish. J Neurophysiol 30(6):1377–1403

Galambos R (1956) Suppression of auditory nerve activity by stimulation of efferent fibers to cochlea. J Neurophysiol 19(5):424–437

Gay LA, Stanfield PR (1977) Cs(+) causes a voltage-dependent block of inward K currents in resting skeletal muscle fibres. Nature 267(5607):169–170

Géléoc GS, Lennan GW, Richardson GP, Kros CJ (1997) A quantitative comparison of mechanoelectrical transduction in vestibular and auditory hair cells of neonatal mice. Proc Biol Sci 264(1381):611–621

Gill SS, Salt AN (1997) Quantitative differences in endolymphatic calcium and endocochlear potential between pigmented and albino guinea pigs. Hear Res 113(1–2):191–197

Glowatzki E, Fuchs PA (2002) Transmitter release at the hair cell ribbon synapse. Nat Neurosci 5(2):147–154
Goldberg JM, Fernández C (1980) Efferent vestibular system in the squirrel monkey: anatomical location and influence on afferent activity. J Neurophysiol 43(4):986–1025
Goldman DE (1943) Potential, impedance, and rectification in membranes. J Gen Physiol 27(1):37–60
Hackney CM, Furness DN, Benos DJ, Woodley JF, Barratt J (1992) Putative immunolocalization of the mechanoelectrical transduction channels in mammalian cochlear hair cells. Proc Biol Sci 248(1323):215–221
Hacohen N, Assad JA, Smith WJ, Corey DP (1989) Regulation of tension on hair-cell transduction channels: displacement and calcium dependence. J Neurosci 9(11):3988–3997
Hagiwara S, Ohmori H (1982) Studies of calcium channels in rat clonal pituitary cells with patch electrode voltage clamp. J Physiol 331:231–252
Hagiwara S, Takahashi K (1974) The anomalous rectification and cation selectivity of the membrane of a starfish egg cell. J Membr Biol 18(1):61–80
Hillman DE (1972) Observations on morphological features and mechanical properties of the peripheral vestibular receptor system in the frog. Porg Brain Res 37:69–75
Hodgkin AL, Katz B (1949) The effect of sodium ions on the electrical activity of giant axon of the squid. J Physiol 108(1):37–77
Holt JR, Stauffer EA, Abraham D, Géléoc GS (2007) Dominant-negative inhibition of M-like potassium conductances in hair cells of the mouse inner ear. J Neurosci 27(33):8940–8951
Hossain WA, Antic SD, Yang Y, Rasband MN, Morest DK (2005) Where is the spike generator of the cochlear nerve? Voltage-gated sodium channels in the mouse cochlea. J Neurosci 25(29):6857–6868
Housley GD, Morton-Jones R, Vlajkovic SM, Telang RS, Paramananthasivam V, Tadros SF, Wong AC, Froud KE, Cederholm JM, Sivakumaran Y, Snguanwongchai P, Khakh BS, Cockayne DA, Thorne PR, Ryan AF (2013) ATP-gated ion channels mediate adaptation to elevated sound levels. Proc Natl Acad Sci U S A 110(18):7494–7499
Hudspeth AJ (1982) Extracellular current flow and the site of transduction by vertebrate hair cells. J Neurosci 2(1):1–10
Hudspeth AJ, Corey DP (1977) Sensitivity, polarity, and conductance change in the response of vertebrate hair cells to controlled mechanical stimuli. Proc Natl Acad Sci U S A 74(6):2407–2411
Jaramillo F, Hudspeth AJ (1991) Localization of the hair cell's transduction channels at the hair bundle's top by iontophoretic application of a channel blocker. Neuron 7(3):409–420
Jorgensen F (1978) Effect of amiloride on the mechanosensitivity of hair cells in the lateral line system of Necturus mac. Acta physiologica Scandinavica abstracts for Scandinavian Physiological Society Meeting (Odense), November 1978, 25-26A
Jorgensen F, Ohmori H (1988) Amiloride blocks the mechano-electrical transduction channel of hair cells of the chick. J Physiol 403:577–588
Kandler K, Clause A, Noh J (2009) Tonotopic reorganization of developing auditory brainstem circuits. Nat Neurosci 12(6):711–717
Kataoka Y, Ohmori H (1994) Activation of glutamate receptors in response to membrane depolarization of hair cells isolated from chick cochlea. J Physiol 477(Pt 3):403–414
Kataoka Y, Ohmori H (1996) Of known neurotransmitters, glutamate is the most likely to be released from chick cochlear hair cells. J Neurophysiol 76(3):1870–1879
Kawashima Y, Géléoc GSG, Kurima K, Labay V, Lelli A, Asai Y, Makishima T, Wu DK, Santina CCD, Holt JR, Griffith AJ (2011) Mechanotransduction in mouse inner ear hair cells require transmembrane channel–like genes. J Clin Invest 121(12):4796–4809
Kawashima Y, Kurima K, Pan B, Griffith AJ, Holt JR (2015) Transmembrane channel-like (TMC) genes are required for auditory and vestibular mechanosensation. Pflugers Arch 467(1):85–94
Kilić G, Sciancalepore M, Cherubini E (1992) Single-channel currents of NMDA type activated by L- and D-homocysteic acid in cerebellar granule cells in culture. Neurosci Lett 141(2):231–235

Kimitsuki T, Ohmori H (1992) The effect of caged calcium release on the adaptation of the transduction current in chick hair cells. J Physiol 458:27–40
Kimitsuki T, Ohmori H (1993) Dihydrostreptomycin modifies adaptation and blocks the mechanoelectric transducer in chick cochlear hair cells. Brain Res 624(1–2):143–150
Kitajiri M, Yamashita T, Tohyama Y, Kumazawa T, Takeda N, Kawasaki Y, Matsunaga T, Girgis S, Hillyard CJ, MacIntyre I, Emson PC, ShiosakaS TM (1985) Localization of calcitonin gene-related peptide in the organ of Corti of the rat: an immunohistochemical study. Brain Res 358(1–2):394–397
Kroese AB, van den Bercken J (1982) Effects of ototoxic antibiotics on sensory hair cell functioning. Hear Res 6(2):183–197
Kroese AB, Das A, Hudspeth AJ (1989) Blockage of the transduction channels of hair cells in the bullfrog's sacculus by aminoglycoside antibiotics. Hear Res 37(3):203–217
Kurima K, Peters LM, Yang Y, Riazuddin S, Ahmed ZM, Naz S, Arnaud D, Drury S, Mo J, Makishima T, Ghosh M, Menon PS, Deshmukh D, Oddoux C, Ostrer H, Khan S, Riazuddin S, Deininger PL, Hampton LL, Sullivan SL, Battey JF Jr, Keats BJ, Wilcox ER, Friedman TB, Griffith AJ (2002) Dominant and recessive deafness caused by mutations of a novel gene, TMC1, required for cochlear hair-cell function. Nat Genet 30(3):277–284
Kurima K, Yang Y, Sorber K, Griffith AJ (2003) Characterization of the transmembrane channel-like (TMC) gene family: functional clues from hearing loss and epidermodysplasia verruciformis. Genomics 82(3):300–308
Kurima K, Ebrahim S, Pan B, Sedlacek M, Sengupta P, Millis BA, Cui R, Nakanishi H, Fujikawa T, Kawashima Y, Choi BY, Monahan K, Holt JR, Griffith AJ, Kachar B (2015) TMC1 and TMC2 localize at the site of Mechanotransduction in mammalian inner ear hair cell Stereocilia. Cell Rep 12(10):1606–1617
Leake PA, Hradek GT, Chair L, Snyder RL (2006) Neonatal deafness results in degraded topographic specificity of auditory nerve projections to the cochlear nucleus in cats. J Comp Neurol 497(1):13–31
Lewis RS (1982) Characterization of voltage- and ion-dependent conductances in vertebrate hair cells. Soc Neurosci Abstr 8:728
Lewis RS, Hudspeth AJ (1983) Voltage- and ion-dependent conductances in solitary vertebrate hair cells. Nature 304(5926):538–541
Llano I, Marty A, Armstrong CM, Konnerth A (1991) Synaptic- and agonist-induced excitatory currents of Purkinje cells in rat cerebellar slices. J Physiol 434:183–213
Llinás R, Steinberg IZ, Walton K (1981) Presynaptic calcium currents in squid giant synapse. Biophys J 33(3):289–321
Marcotti W, Johnson SL, Rusch A, Kros CJ (2003) Sodium and calcium currents shape action potentials in immature mouse inner hair cells. J Physiol 552(Pt 3):743–761
Masetto S, Bosica M, Correia MJ, Ottersen OP, Zucca G, Perin P, Valli P (2003) Na^+ currents in vestibular type I and type II hair cells of the embryo and adult chicken. J Neurophysiol 90(2):1266–1278
Matsuura S, Ikeda K, Furukawa T (1971) Effects of streptomycin, kanamycin, quinine, and other drugs on the microphonic potentials of goldfish sacculus. Jpn J Physiol 21(5):579–590
Mayer ML, Westbrook GL (1987) The physiology of excitatory amino acids in the vertebrate central nervous system. Prog Neurobiol 28(3):197–276
McBain CJ, Mayer ML (1994) N-methyl-D-aspartic acid receptor structure and function. Physiol Rev 74(3):723–760
McKay SM, Oleskevich S (2007) The role of spontaneous activity in development of the endbulb of held synapse. Hear Res 230(1–2):53–63
Meyer J, Furness DN, Zenner HP, Hackney CM, Gummer AW (1998) Evidence for opening of hair-cell transducer channels after tip-link loss. J Neurosci 18(17):6748–6756
Nakagawa T, Akaike N, Kimitsuki T, Komune S, Arima T (1990) ATP-induced current in isolated outer hair cells of guinea pig cochlea. J Neurophysiol 63(5):1068–1074

New HV, Mudge AW (1986) Calcitonin gene-related peptide regulates muscle acetylcholine receptor synthesis. Nature 323(6091):809–811
Nowak L, Bregestovski P, Ascher P, Herbet A, Prochiantz A (1984) Magnesium gates glutamate-activated channels in mouse central neurons. Nature 307(5950):462–465
Ohmori H (1978) Inactivation kinetics and steady-state current noise in the anomalous rectifier of tunicate egg cell membranes. J Physiol 281:77–99
Ohmori H (1984a) Mechanoelectrical transducer has discrete conductances in the chick vestibular hair cell. Proc Natl Acad Sci U S A 81(6):1888–1891
Ohmori H (1984b) Studies of ionic currents in the isolated vestibular hair cell of the chick. J Physiol 350:561–581
Ohmori H (1985) Mechano-electrical transduction currents in isolated vestibular hair cells of the chick. J Physiol 359:189–217
Ohmori H (1987) Gating properties of the mechano-electrical transducer channel in the dissociated vestibular hair cell of the chick. J Physiol 387:589–609
Ohmori H (1988) Mechanical stimulation and Fura-2 fluorescence in the hair bundle of dissociated hair cells of the chick. J Physiol 399:115–137
Oliver D, Plinkert P, Zenner HP, Ruppersberg JP (1997) Sodium current expression during postnatal development of rat outer hair cells. Pflugers Arch 434(6):772–778
Pan B, Géléoc GS, Asai Y, Horwitz GC, Kurima K, Ishikawa K, Kawashima Y, Griffith AJ, Holt JR (2013) TMC1 and TMC2 are components of the mechanotransduction channel in hair cells of the mammalian inner ear. Neuron 79(3):504–515
Pickles JO (1993) A model for the mechanics of the stereocilia bundle on acousticolateral hair cells. Hear Res 68(2):159–172
Pickles JO, Comis SD, Osborne MP (1984) Cross-links between stereocilia in the guinea pig organ of Corti, and their possible relation to sensory transduction. Hear Res 15(2):103–112
Ricci AJ, Fettiplace R (1997) The effects of calcium buffering and cyclic AMP on mechano-electrical transduction in turtle auditory hair cells. J Physiol 501(Pt 1):111–124
Rosenmund C, Legendre P, Westbrook GL (1992) Expression of NMDA channels on cerebellar Purkinje cells acutely dissociated from newborn rats. J Neurophysiol 68(5):1901–1905
Rubel EW, Fritzsch B (2002) Auditory system development: primary auditory neurons and their targets. Annu Rev Neurosci 25:51–101
Rüsch A, Kros CJ, Richardson GP (1994) Block by amiloride and its derivatives of mechano-electrical transduction in outer hair cells of mouse cochlear cultures. J Physiol 474(1):75–86
Russell IJ, Sellick PM (1978) Intracellular studies of hair cells in the mammalian cochlea. J Physiol 284:261–290
Russell IJ, Sellick PM (1983) Low-frequency characteristics of intracellularly recorded receptor potentials in guinea-pig cochlear hair cells. J Physiol 338:179–206
Sand O (1975) Effects of different ionic environments on the mechanosensitivity of lateral line organs in the mudpuppy. J Comp Physiol A 102:27–42
Sariban-Sohraby S, Benos DJ (1986) The amiloride-sensitive sodium channel. Am J Phys 250(2 Pt 1):C175–C190
Sewell WF, Starr PA (1991) Effects of calcitonin gene-related peptide and efferent nerve stimulation on afferent transmission in the lateral line organ. J Neurophysiol 65(5):1158–1169
Sewell WF, Norris CH, Tachibana M, Guth PS (1978) Detection of an auditory nerve--activating substance. Science 202(4370):910–912
Shigemoto T, Ohmori H (1990) Muscarinic agonists and ATP increase the intracellular Ca2+ concentration in chick cochlear hair cells. J Physiol 420:127–148
Shigemoto T, Ohmori H (1991) Muscarinic receptor hyperpolarizes cochlear hair cells of chick by activating ca(2+)-activated K+ channels. J Physiol 442:669–690
Spangler KM, Warr WB (1991) The descending auditory system. In: Altschuler RA, Bobbin RP, Clopton BM, Hoffman DW (eds) Neurobiology of hearing: the central auditory system. Raven, New York, pp 27–45
Spoendlin H (1972) Innervation densities of the cochlea. Acta Otolaryngol 73(2):235–248

Sridhar TS, Liberman MC, Brown MC, Sewell WF (1995) A novel cholinergic "slow effect" of efferent stimulation on cochlear potentials in the guinea pig. J Neurosci 15(5 Pt 1):3667–3678
Standen NB, Stanfield PR (1978) A potential- and time-dependent blockade of inward rectification in frog skeletal muscle fibres by barium and strontium ions. J Physiol 280:169–191
Standen NB, Stanfield PR (1979) Potassium depletion and sodium block of potassium currents under hyperpolarization in frog sartorius muscle. J Physiol 294:497–520
Steel KP, Barkway C (1989) Another role for melanocytes: their importance for normal stria vascularis development in the mammalian inner ear. Development 107(3):453–463
Takasaka T, Smith CA (1971) The structure and innervation of the pigeon's basilar papilla. J Ultrastruct Res 35(1):20–65
Takeda N, Kitajiri M, Girgis S, Hillyard CJ, MacIntyre I, Emson PC, Shiosaka S, Tohyama M, Matsunaga T (1986) The presence of a calcitonin gene-related peptide in the olivocochlear bundle in rat. Exp Brain Res 61(3):575–578
Tan X, Beurg M, Hackney C, Mahendrasingam S, Fettiplace R (2013) Electrical tuning and transduction in short hair cells of the chicken auditory papilla. J Neurophysiol 109(8):2007–2020
Tanaka K, Smith CA (1978) Structure of the chicken's inner ear: SEM and TEM study. Am J Anat 153(2):251–271
Tasaki I, Davis H, Eldredge DH (1954) Exploration of cochlear potentials in Guinea pig with a microelectrode. J Acoust Soc Am 26:765–773
Thornhill R (1972) The effect of catecholamine precursors and related drugs on the morphology of the synaptic bars in the vestibular epithelia of the frog, Rana temporaria. Comp Gen Pharmacol 3:89–97
Tilney LG, Saunders JC (1983) Actin filaments, stereocilia, and hair cells of the bird cochlea. I. Length, number, width, and distribution of stereocilia of each hair cell are related to the position of the hair cell on the cochlea. J Cell Biol 96(3):807–821
Turner RG, Muraski AA, Nielsen DW (1981) Cilium length: influence on neural tonotopic organization. Science 213(4515):1519–1521
van Netten SM, Kros CJ (2007) Insights into the pore of the hair cell transducer channel from experiments with permeant blockers. Curr Top Membr 59:375–398
Vollrath MA, Kwan KY, Corey DP (2007) The micromachinery of mechanotransduction in hair cells. Annu Rev Neurosci 30:339–365
Vreugde S, Erven A, Kros CJ, Marcotti W, Fuchs H, Kurima K, Wilcox ER, Friedman TB, Griffith AJ, Balling R, Hrabé De Angelis M, Avraham KB, Steel KP (2002) Beethoven, a mouse model for dominant, progressive hearing loss DFNA36. Nat Genet 30(3):257–258
Warr WB (1992) Organization of olivocochlear efferent systems in mammals. In: Webster DB, Popper AN, Fay RR (eds) The mammalian auditory pathway: neuroanatomy. Springer, New York, pp 410–448
Webster DB (1992) An overview of mammalian auditory pathways with an emphasis on humans. In: Webster DB, Popper AN, Fay RR (eds) The mammalian auditory pathways: neuroanatomy. Springer, New York, pp 1–22
Wersaell J, Flock A, Lundquist P-G (1965) Structural basis for directional sensitivity in cochlear and vestibular sensory receptors. Cold Spring Harb Symp Quant Biol 30:115–132
Williams DA, Fogarty KE, Tsien RY, Fay FS (1985) Calcium gradients in single smooth muscle cells revealed by the digital imaging microscope using Fura-2. Nature 318(6046):558–561
Yamaguchi K, Ohmori H (1990) Voltage-gated and chemically gated ionic channels in the cultured cochlear ganglion neuron of the chick. J Physiol 420:185–206
Yamaguchi K, Ohmori H (1993) Suppression of the slow K+ current by cholinergic agonists in cultured chick cochlear ganglion neurons. J Physiol 464:213–228
Yoshida N, Shigemoto T, Sugai T, Ohmori H (1994) The role of inositol trisphosphate on ACh-induced outward currents in bullfrog saccular hair cells. Brain Res 644(1):90–100

Chapter 2
Signal Processing in the Brainstem Auditory Nuclei

ANFs bifurcate in the cochlear nucleus, and one branch innervates cells in anteroventral cochlear nucleus (AVCN), and the other branch innervates cells in posteroventral cochlear nucleus (PVCN) and dorsal cochlear nucleus (DCN) in mammals. Auditory features are aligned in each nucleus in the order of characteristic frequency (CF), or the arrangement is called tonotopic organization. This map of frequency responsibility reflects original frequency responsibility in the cochlea organ.

Avian ANFs also bifurcate right after entering the brainstem and innervates two subnuclei of the cochlear nucleus (Figs. 2.1 and 2.2). One branch of the ANFs makes synaptic connections with neurons in the nucleus magnocellularis (NM), and the other branch makes synapses on neurons in the nucleus angularis (NA). The temporal information and level information of sound are extracted separately in these two nuclei. This distinct encoding pathway was first reported in the barn owl cochlear nucleus (Takahashi and Konishi 1988; Sullivan and Konishi 1984; Takahashi et al. 1984), where neurons in the NM responded to a particular phase of sinusoidal sound, while the spike count was saturated at the level slightly above the threshold intensity of sound. In the NA, neurons respond to sinusoidal sound without phase sensitivity, although responses of NA neurons had a large dynamic range to the sound intensity. Accordingly, NM neurons are considered to respond to the temporal information and send the signal to neurons in the nucleus laminaris (NL) of both sides (Fig. 2.9), while neurons in NA encode and relay the sound level information to the nucleus lateral lemniscus (LL) of contralateral side (Fig. 2.1, 2.38). In the NL, binaural temporal signals are processed to encode the interaural time difference (ITD), which is the difference of sound arrival time between two ears. ITD is an important cue for the sound source localization, particularly in the horizontal plane. In the LL, the difference of sound level information (ILD, interaural level difference) is processed. In the barn owl, ILD is used as a cue for sound source localization in the vertical axis. Different feature extractions in these two nuclei were demonstrated in the barn owl by injection of lidocaine (a local anesthetic). The corresponding effects were observed on the firing property of space-specific neurons in the IC (inferior colliculus or MLd in birds, mesencephalicus lateralis dorsa-

H. Ohmori, *Auditory Information Processing*,
https://doi.org/10.1007/978-981-32-9713-5_2

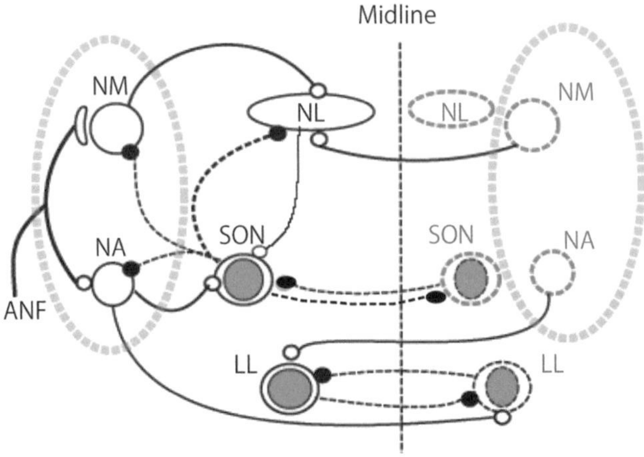

Fig. 2.1 A scheme of avian brainstem auditory nuclei
ANF bifurcates and innervates NM and NA. The pathway starting from NM processes the temporal information of sound, and the pathway from NA processes the sound level information. SON receives sound level signal from NA and sends inhibitory fibers to NL, NM, and NA and contralateral SON. LL mutually inhibits contralateral LL.
ANF auditory nerve fiber, *NM* nucleus magnocellularis, *NA* nucleus angularis, *NL* nucleus laminaris, *LL* lateral lemniscus, *SON* superior olivary nucleus

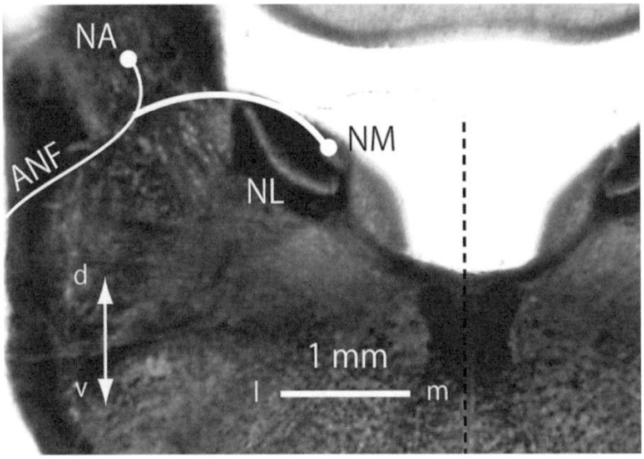

Fig. 2.2 A slice of chick brainstem
NA, NM, and NL are included in the slice. The pattern of ANF innervation is illustrated. Midline is indicated by a broken line

lis) (Takahashi et al. 1984). Injection of lidocaine to the NM reduced the sensitivity to ITD, while the sensitivity to ILD was intact in IC neurons. Alternatively, the injection to the NA reduced the sensitivity to ILD but left the sensitivity to ITD intact. It is of great interest to understand how these different responses to sound features are generated in these two nuclei while receiving basically the same set of signals as ANFs activity.

2.1 Neural Activity in NM (Nucleus Magnocellularis)

Neurons in NM generate a single action potential at the onset of current injection or auditory nerve stimulation in slice preparations (Zhang and Trussell 1994; Koyano et al. 1996). Size of EPSC (excitatory postsynaptic current) was large, kinetics was fast, and the current was mediated by calcium permeable AMPA-type glutamate receptors (Raman and Trussell 1992; Otis et al. 1995; Trussell 1997; Levin et al. 1997). The flop splice variants of GluR4 subunit, GluR4-flop, and a reduced content of GluR2 subunit may mediate the temporally accurate responses of NM neurons to the auditory nerve stimulation (Raman et al. 1994; Gardner et al. 2001). GluR4-flop subunit quickly gates the channel, while reduced level of GluR2 makes calcium ions permeable through the AMPA receptor channels. GluR2 subunits coexpressed with other subunits will suppress the calcium permeability. Size of EPSC in NM was, however, different along the frequency regions of the nucleus (Fukui and Ohmori 2004). In the low-frequency region of NM, EPSC was small, and the size was increased with the stimulus intensity, while in the middle to high-frequency regions, EPSC was large and was generated in all or none manner in response to auditory nerve stimulation. We separated the frequency region roughly into three; low (lower than 500 Hz), middle (500–2500 Hz), and the high (higher than 2500 Hz) based on the topographical organization of the NM (Rubel and Parks 1975). Along with the tonotopy, membrane excitability of NM neurons and morphology of synaptic terminals were different, which will be described in details below.

2.1.1 Tonotopic Expression of Kv1 Channels

Spike firing of NM neurons was affected by both low-voltage-activated (KLVA) and high-voltage-activated (KHVA) K^+ currents (Reyes et al. 1994; Koyano et al. 1996; Rathouz and Trussell 1998; Trussell 1999; Parameshwaran et al. 2001). Block of KLVA currents by DTX (dendrotoxin, a blocker of Kv1 channels) generated repetitive firing, while the block of KHVA by TEA (tetraethyl ammonium) broadened action potentials.

Membrane excitability of NM neurons was different across the frequency regions of the nucleus. The expression level of low-voltage-activated (KLVA) K^+ channels determined the major difference (Rathouz and Trussell 1998). Tonotopic gradients were found in various aspects of NM neurons (Fig. 2.3). In high-frequency neurons, resting potential was more negative, input resistance and the membrane time constant were smaller, spike threshold current was larger, and the threshold potential was higher than neurons in the middle to low-frequency NM regions. All these parameters were sensitive to DTX, and the DTX effects were largest in the high-frequency neurons (Fukui and Ohmori 2004). After bath application of DTX (Fig. 2.3), resting potential became more positive, and threshold potential became

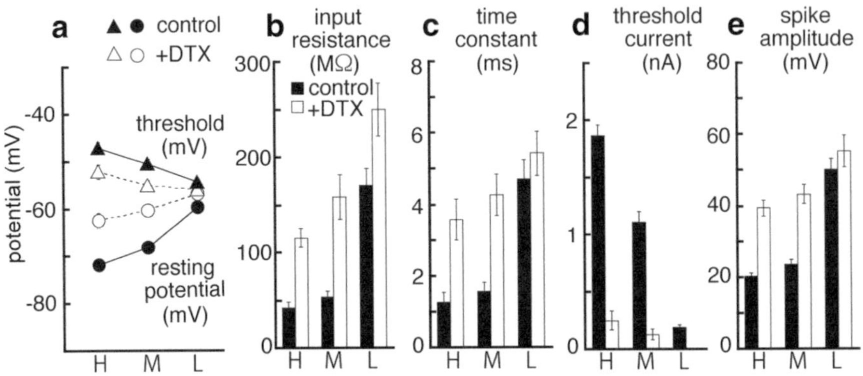

Fig. 2.3 CF dependence of membrane excitability parameters of NM neurons in slices (**a**) resting potential. (**b**) input impedance. (**c**) membrane time constant. (**d**) threshold current. (**e**) spike amplitude. Effects of DTX (100 nM) are indicated by open bars. H, M, and L indicate high, middle, and low-CF NM neurons. *CF* characteristic frequency
From Fukui and Ohmori (2004)

lower, or threshold current was decreased. Since DTX is a blocker of KLVA, most properties of NM neurons are determined by the expression gradient of KLVA. Multiple spikes were generated by current injection after application of DTX (Fukui and Ohmori 2004).

DTX blocks the low-voltage-activated (KLVA) K^+ channels such as Kv1.1, Kv1.2, or Kv1.6 (Hopkins et al. 1994; Harvey 2001). Kv1.1 mRNA was expressed with a tonotopic gradient in NM, higher toward high-frequency region. Kv1.2 mRNA was positive, but the expression was not different between frequency regions. Moreover, antibody to Kv1.2 did not show any immunological reaction products. The immunoreactivity level of Kv1.1 was higher in the high- to middle-frequency region than in the low-frequency region of the nucleus (Fukui and Ohmori 2004). Accordingly, the tonotopic difference of excitability of NM neurons is based on the expression level of Kv1.1 channel.

2.1.2 EPSCs

EPSCs recorded in neurons of high- to middle-frequency NM regions were large and generated mostly in all-or-nothing manner when electrical stimulation was applied to the ANFs in slice preparations (Fig. 2.4). In contrast, in the low-frequency NM region, EPSC was small, and the size increased gradually when the stimulus intensity was increased. Temporal jitter of spikes was small regardless of NM frequency regions when ANFs were electrically stimulated; 9 μsec in high vs 21 μsec in low-frequency cells as measured as the standard deviation of spike timing. These temporal jitters were much larger in NM neurons in vivo when acoustic stimulation

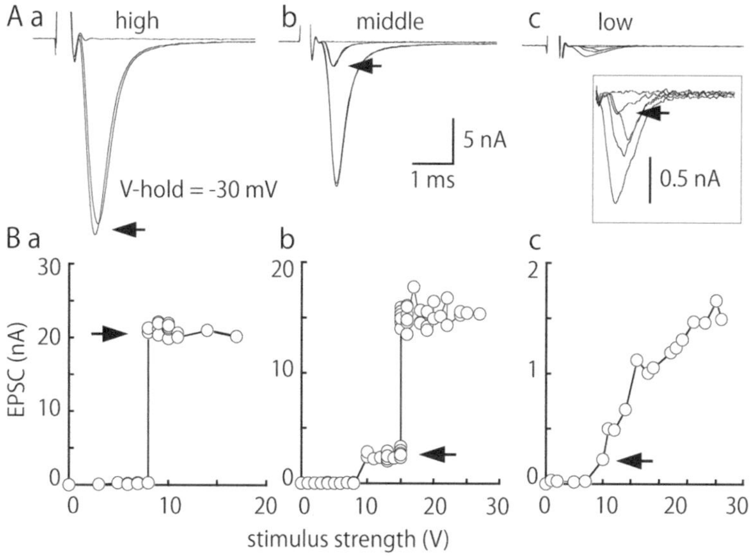

Fig. 2.4 Synaptic currents in NM neurons
A, synaptic currents in high-CF (Aa), middle-CF (Ab), and low-CF (Ac) neurons. Inset in Ac illustrates EPSCs of a low-CF neuron in expanded scale. B, in high (Ba) or middle (Bb) CF NM neurons, EPSCs were large and generated in all or nothing manner. In low-CF neurons (Bc), EPSCs of small size were generated, and the size was increased depending on the stimulus intensity. Arrows indicate the smallest EPSC
From Fukui and Ohmori (2004)

was applied, as it will be described later (Fig. 2.6, Fukui et al. 2006). This might be because the jitter of spike timings in NM in vivo included the jitter of spike timings of ANFs (Fig. 2.6). Responsibility of neurons to high-frequency stimulus was not different in CF regions. All neurons followed the stimulus rate up to 333 Hz when ANFs were stimulated electrically in slices; 333 Hz was the highest frequency applied by our electrical stimulation.

2.1.3 Tonotopic Difference of Synapse Terminal Morphology

Synaptic terminals of ANFs have various shapes. The different terminal shape was associated with different features of synaptic transmission. Characteristically, the end bulb of Held, which is one of the largest terminals in the mammalian brain, is formed between ANFs and spherical bushy cell (SBC) or globular bushy cell (GBC) in the cochlear nucleus, and the calyx of Held is formed between the axon terminal of globular bushy cell and neurons in the medial nucleus of the trapezoid body (MNTB). All these neurons have distinct roles in processing phase-sensitive auditory

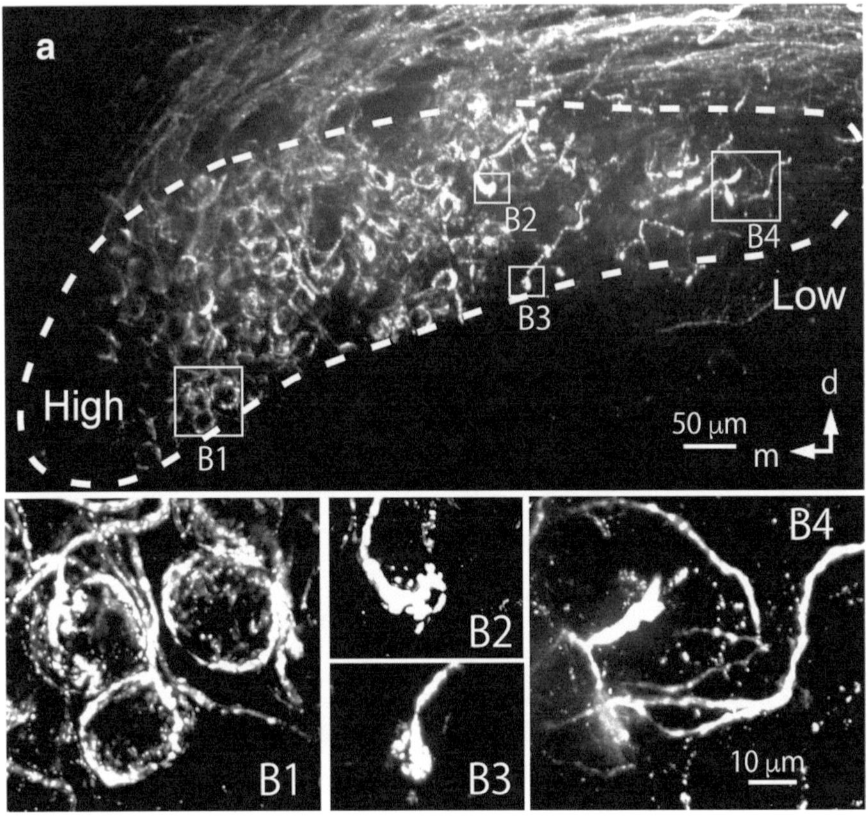

Fig. 2.5 ANF projections to NM
(**a**) ANF projections were visualized by placing DiI crystal into the cochlear organ. The whole image of projections to NM is shown in **a**. High, the high-CF region, and Low, the low-CF region. B1 to B4, ANF terminal structures on NM neurons. B1, the high-CF region; B2 and B3, middle-CF NM region. B4, the low-CF NM region. Calibration in B4 is applied to B1–B4

signals. The high and middle-frequency NM neurons of birds are similarly contacted with a large calyx-shaped synaptic terminals (Fig. 2.5) (Fukui and Ohmori 2004).

We visualized ANF terminals by carbocyanine dye DiI (1,1′-dioctadecyl-3,3,3′,3′-tetramethylindocarbocyanine perchlorate). DiI crystal was placed in the cochlear duct of a chick for 1–2 days, and the terminal structure was visualized in NM slices (Fukui and Ohmori 2004). The calyx-shaped end bulb of Held surrounded the cell soma in the high to middle-frequency NM regions (Parks and Rubel 1978; Jhaveri and Morest 1982; Hackett et al. 1982; Carr and Bourdreau 1991), while only bouton-shaped small terminal structures were found in the low-frequency NM region (Fig. 2.5). This was consistent with the observation of small synaptic currents in the low-frequency NM cells, in contrast to the large synaptic currents in the high to middle-frequency NM neurons (Fig. 2.4). In the barn owl, it was also reported that ANFs form terminal boutons on NM neurons in low-frequency region (Köppl 1994; Köppl and Carr 1997). In the basilar papilla, which is the avian cochlear organ and where ANFs originated, the low-frequency region (lower than 500 Hz) occupies 40–50% of the area from the apex (Ryals and Rubel 1982). However, the area of low-frequency ANFs projected in NM is only 25% (Lippe and Rubel 1985). Moreover, the basilar papilla is wider toward the apex (the low-frequency region), and the hair cell density in the apex is approximately three times higher than that in the base (Ryals and Rubel 1982; Tilney and Saunders 1983). Consistently, the number of ANFs innervating one low-CF NM neuron was approximately five times greater than that of the high-CF NM neurons (Fukui and Ohmori 2004). This is also consistent with the increase of EPSC size with increase of stimulus intensity applied to the ANFs in low-CF NM neurons, in contrast to the all-or-none response property in the high to middle-CF NM neurons in slice experiments (Fig. 2.4). ANF diameter was thick in the middle to high-CF region of NM (1.8–2.1 μm) and thin in the low-CF region (1.1–1.3 μm).

2.1.4 Reduced Timing Jitter of Spike Firing in the Low-Frequency NM Cells

A large EPSC or EPSP is advantageous to make a high-fidelity transmission, which is the synaptic transmission that reflects faithfully the timing of input signal. Input from a single ANF was large and sufficient to generate an action potential in the high to middle-CF NM neurons. In contrast, the small EPSC or EPSP of low-CF NM neurons was too small to reach the threshold by its own and requires summation of a number of synaptic inputs to fire a spike (Fig. 2.4). This appeared disadvantageous for auditory signal transmission because of the extra time needed for summation; however, the summation of smaller EPSPs was demonstrated effective in reducing timing jitter of spike firing activity (Joris et al. 1994; Xu-Friedman and Regehr 2005; Fukui et al. 2006). The phase-locked firing property was better in NM neurons than in the ANFs of similar low CF (Fig. 2.6). This phenomenon of phase improvement during synaptic transmission was due to the elimination of out-of-phase synaptic inputs during summation of EPSPs and was first reported in bushy cells of the cat cochlear nucleus, the mammalian homologue of NM neurons (Joris et al. 1994).

The spike firing was phase-locked to sound input of low frequency in both ANFs and NM neurons in the chick, when tonal sound stimulation was applied (around 370 Hz in Fig. 2.6a, b). These spikes of ANFs and NM neurons were recorded by separate penetrations of electrode. Interestingly, the spike firing was phase-locked much greater extent in NM unit than in ANF unit; the maximum vector strength was

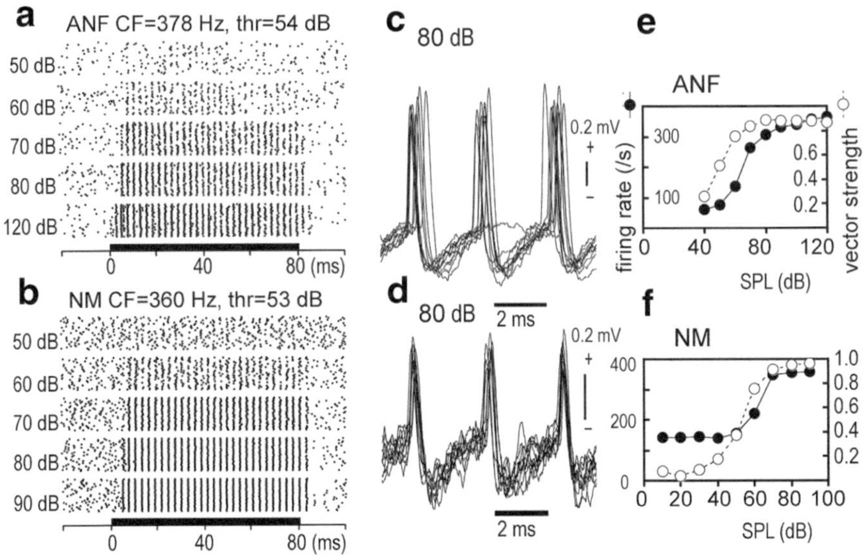

Fig. 2.6 Sound-evoked unit spikes
(**a**) dot raster plots of 378 Hz BF ANF unit. Sound level was increased from 50 dB at the top to 120 dB at the bottom. 80 ms sound exposure time is indicated by the bar at the bottom in **a** and **b**. Ten trials were plotted at each sound pressure level (SPL). (**b**) NM unit of 360 Hz BF. Dot raster plots. Sound level was from 50 dB to 90 dB. Note that timing fluctuation is smaller in the NM unit than the ANF unit of similar BF. thr; threshold SPL. (**c** and **d**) spikes of ANF and NM units were overlaid, which responded to 80 dB tonal sound of CF. Note a smaller fluctuation of peak time in NM unit than in ANF unit. (**e** and **f**) SPL dependence of firing rate (filled circles, left ordinate) and vector strength (open circles, right ordinate), from ANF unit (**e**) and NM unit (**b**)
From Fukui and Ohmori (2004)

0.89 for ANF unit and 0.95 for NM unit. Vector strength indicates the degree of phase-locked firings of neurons and is 1 when all spike activity occurred totally in-phase and 0 when totally out-of-phase (Goldberg and Brown 1969). The phase-locked firing was improved with the increase of stimulus sound pressure level (Fig. 2.6e, f, open circles). The vector strength was not different between ANFs and NM units at the sound level of rate threshold and reached nearly a plateau level around 20 dB higher than the rate threshold in most units (Fukui et al. 2006).

The temporal jitter was particularly small in NM neurons when the sound intensity was high (90 dB, Fig. 2.6b). Spike timing jitter, defined as the standard deviation (s.d.) of spike timing, was sound frequency dependent and decreased from 300 μs at the low frequency to 100 μs at the high frequency, both in ANF units and NM units (Köppl 1997; Fukui et al. 2006). Moreover, in the low-frequency NM neurons, the spike timing jitter was smaller than that of ANFs by about 50 μs.

Accordingly, the temporal precision of spike activity to low-frequency sound was improved during transmission across the synapse between ANF and NM.

Spontaneous firing rate of NM units (100–150 Hz, Fig. 2.6f, filled circles) was higher than that of ANF units (50–60 Hz, Fig. 2.6e, filled circles). The rate threshold was nearly the same, and the maximum firing rate (about 400 Hz) was highest in the middle-frequency region (CF 0.8–1 kHz) in both units (Fukui et al. 2006). These features were consistent with the report in other avian species (Chicken, Warchol and Dallos 1990, Manley et al. 1991, Salvi et al. 1999, Saunders et al. 2002; Barn Owl, Köppl 1997).

We made a simulation using a *NEURON* model to understand the advantage of having multiple small EPSCs in the transmission of low-frequency ANF-NM synapses. A multiple-compartment model of NM neurons was made, and two types of ANF synapses were assigned to the model neuron inputs: (1) a single synapse input of a large conductance (44 nS) or (2) multiple synapse inputs of a small conductance (16 inputs, 4 nS) (Fig. 2.7). The timing of individual inputs was randomly varied with a s.d. of 300 μs in each sound cycle (a jitter mimicked that observed in the low-frequency ANFs in vivo, Fukui et al. 2006). In case of a large single synapse input in the model neuron, spikes were generated following precisely the input timing (Fig. 2.7Ac). The jitter of spike output, s.d. of spike peak time (Fig. 2.7Ac) was the same (300 μs) as that of input ANF activity (Fig. 2.7Ab). When aligned at the onset timing of the EPSP, spikes were generated at a uniform timing with a short delay (500 μs, Fig. 2.7Ad). However, in the case of multiple synapse inputs of small conductance, the spike occurred after summation of EPSPs and in a more compact timing (Fig. 2.7Bc). The jitter of spike outputs was far smaller (140 μs) than the jitter of input ANF (300 μs, Fig. 2.7Bb). When these spikes were aligned at the onset timing of the first EPSP, it is clear that spikes were generated with a longer delay (average delay 860 μs, Fig. 2.7Bd). During the slow rise of summated EPSP, inactivation of Na^+ channels progressed, and the spike threshold was increased. These drawbacks in spike generation were compensated for by the accumulation of Na^+ channels in the axon initial segment of low-frequency NM neurons (Fig. 2.8, Kuba and Ohmori 2009) and also by the smaller current of KLVA channel in these neurons (Fukui and Ohmori 2004).

Accordingly, the low-frequency NM neurons are not simply the relay neurons but are the coincidence detector of monaural synaptic inputs. In contrast, the high-frequency NM neurons receive small number of inputs, each having a large quantal content, and generate a high-fidelity one-to-one synaptic transmission and preserve the spike timing information of ANFs activity (Trussell 1999; Fukui and Ohmori 2004).

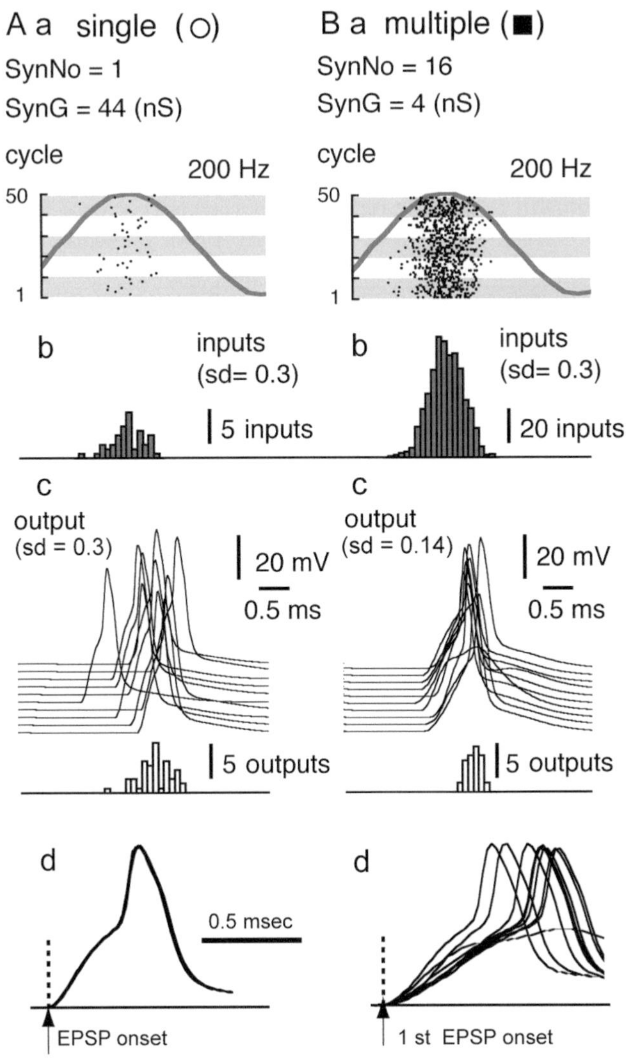

Fig. 2.7 Simulation of NM unit firing activity using a neuron model
(A) NM neuron receives a single ANF synapse of a large conductance (44 nS). (Aa) dot raster plots of 50 trials when sinusoidal sound stimulus (200 Hz, overlaid gray line) was applied. Spikes were generated with temporal jitter of 0.3 (ms, s.d.). (Ab) PSTH (peri-stimulus time histogram) of 50 spike trials in (Aa). (Ac) 10 traces of NM spikes. Bottom is the PSTH of output spikes. (Ad) Spike traces in (Ac) are aligned at the EPSP onset. Every spike followed a same time course. (B) NM neuron receives 16 ANF synapses of small conductance (4 nS). (Ba)–(Bd) as same as corresponding panels in A. (Bc) Timing of spike output is more compact here than in A. (Bd) Spike traces in (Bc) are aligned at the first EPSP onset. The delay to spike generation was fluctuating, and the mean delay time was longer here than in A
From Kuba and Ohmori (2009)

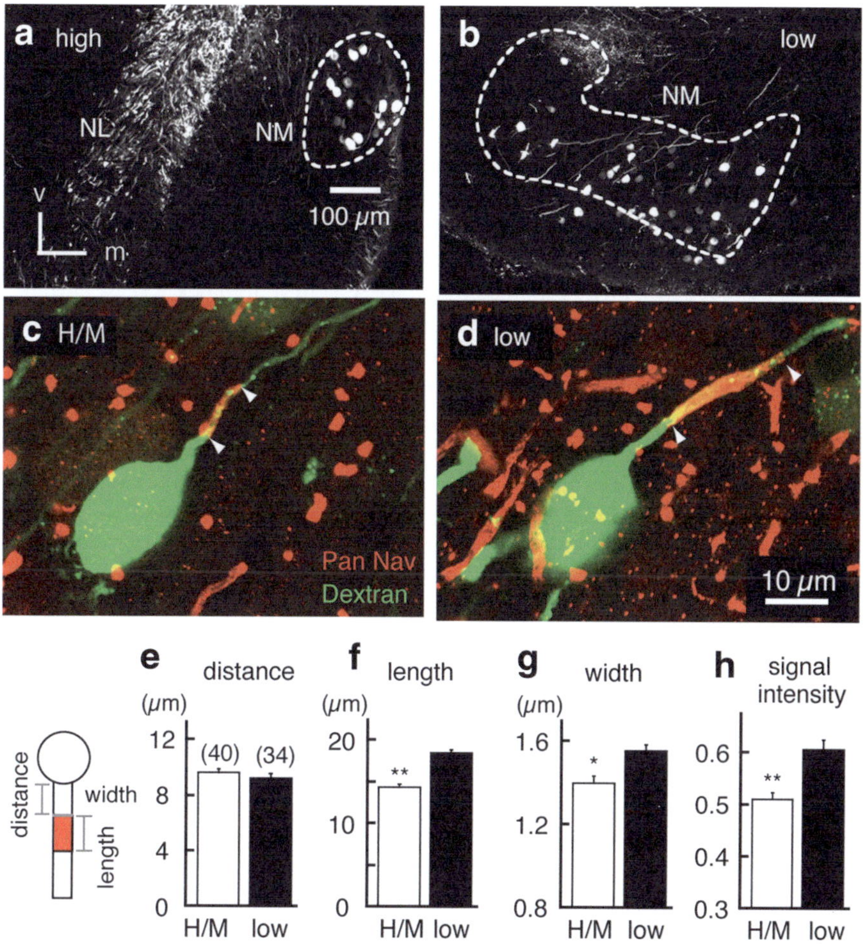

Fig. 2.8 Na channel immunostaining in NM
(**a**, **b**) NM neurons were retrogradely labeled by dextran-Alexa488 injected in the contralateral NL in P3–P6 chicks. Sections were made 2 days after the injection. **a** is the high-CF region, and **b** is the low-CF region. (**c** and **d**) Na channels were immuno-labeled (Pan Nav), and the axon and soma of NM neurons were labeled by Alexa488. AIS (axon initial segment) is indicated as a cluster of Nav channels (red) on the retrogradely labeled proximal axon (green) and is indicated by a pair of arrow head. **c** is from high to middle-CF region. Dot-like structures indicate nodes of Ranvier, and fiber-like structures indicate the AIS. **d** is from low-CF region. Note AIS is longer in **d** than in **c**. Several parameters of AIS are measured: (**e**) the distance from soma, (**f**) the length, (**g**) the width, and (**h**) the signal intensity. Schema in the left illustrates the definition of each parameter
From Kuba and Ohmori (2009)

2.1.5 Distribution of Na^+ Channel in the NM Axon Initial Segment

Precision of spike timing is also affected by the distribution of voltage-gated Na^+ (Nav) channels in the axon initial segment. When spike sizes were compared in NM neurons in slice preparations, spikes were consistently larger in low-frequency cells than in high or middle-frequency cells. The maximum rate of rise was largest in the low-frequency cells, indicating that Na^+ current was large there. The block of KLVA currents by DTX (80 nM), or application of TEA (5 mM, to block KHVA current), did not affect the spike size. The maximum rate of rise was still largest in the low-frequency neurons. Thus, the Na^+ channel conductance was large in the low-frequency NM neurons (Kuba and Ohmori 2009).

We examined the density and the locus of Nav channel expression in NM neurons with immunohistochemistry using antibody that recognized all Nav channel subtypes (Pan Nav) or the antibody to Nav1.6 (Fig. 2.8, Kuba and Ohmori 2009). Strong immunoreactivity of Nav channels appeared as either fiber-like or small dot-like structures by both Pan Nav and Nav1.6 antibodies in the NM. NM neurons project axons to bilateral nucleus laminaris (NL), and the retrograde tracer (Alexa 488-dextran-10 K) injection in NL labeled axons and somas of NM neurons. Double labeling with Na^+ channels antibody revealed that the fiber-like Nav channel clusters were on the proximal part of the axon of NM neurons, presumably at the axon initial segment (AIS, Fig. 2.8c, d). The structure of axon initial segment was measured and was longest in the low-frequency neurons, and the signal intensity was the highest (Fig. 2.8f, h), but the distance from the cell soma was the same regardless of CFs of the neurons (Fig. 2.8e, about 10 μm). Therefore, more Nav channels were accumulated in the axon initial segment of the low-frequency neurons, which likely coped with the progress of Na^+ channel inactivation during the slow depolarization due to the summation of synaptic potential (Fig. 2.7Bd).

It should be noted here that Na^+ channels were not measurably expressed in the NM neuron soma (Fig. 2.8c, d) but were expressed densely at the axon initial segment (Kuba and Ohmori 2009). As it was discussed already in the cochlear ganglion neurons, spikes were generated in the axon initial segment, and the cell body may not contribute for the spike generation, which possibly decreases the capacitive load of spike firing and accelerates the membrane potential change (Hossain et al. 2005).

2.1.6 Axon Initial Segment Is Plastic

Axon initial segment (AIS) is the proximal unmyelinated compartment of axon and is the site of action potential generation (Araki and Otani 1955; Coombs et al. 1957; Bender and Trussell 2012; Stuart et al. 1997; Clark et al. 2005; Palmer and Stuart 2006; Khaliq and Raman 2006). The molecular organization and the role of scaffolding proteins are discussed in detail (Rasband 2010). AIS is the most effective site to regulate neural activity since it is the site where spikes begin. Na^+ channels are accumulated in high density there, and the threshold was lowest for spike generation (Kole et al. 2008; Kole and Stuart 2008). AIS was considered traditionally a static and uniform structure, but recently AIS was found plastic depending on the neural activity (Kuba et al. 2006; Kuba and Ohmori 2009; Kuba et al. 2010; Grubb and Burrone 2010).

Deprivation of afferent inputs in neural circuits leads to diverse plastic changes in both pre- and postsynaptic elements, which facilitated restoration of original neural function (homeostatic plasticity; Cudmore et al. 2010; Wang et al. 2011). Consistently, deprivation of auditory inputs by surgical removal of cochlea leads to an increase in AIS length, thus augmenting the excitability of NM neurons after the hearing loss (Kuba et al. 2010). The length of AIS, defined by the distribution of voltage-gated Na^+ channels and the anchoring protein (ankyrin-G), increased 1.7 times in 7 days after auditory input deprivation. This was accompanied by an increase in the whole-cell-recorded Na^+ current, thus the increase of membrane excitability. Accordingly, homeostatic regulation of AIS may contribute to restore and maintain the auditory activity. Moreover, it was demonstrated that the spike initiation site was under a regulation of plastic activity of neurons and was a powerful target for refining the neuronal computation. Plastic reformation of AIS was similarly found in the neuron of NL (Kuba et al. 2014). Roles of AIS in auditory coincidence detection in NL will be further described in Sect. 2.8.4 of this chapter.

2.2 Fast EPSP in Brainstem Auditory Neurons

Fast time course of EPSP was attributed to the short membrane time constant, the activation of KLVA current, and the fast desensitization kinetics of glutamate (AMPA) receptors. The desensitization of glutamate response was significantly fast in neurons of auditory nuclei, NM, NA, and NL (Raman et al. 1994). Ionotropic glutamate receptors of the AMPA subtypes are assembled from four protein subunits, GluR1–4 (or A–D in other nomination). As splice variants, there are flip-flop variations in these subunits, and the flop-type receptors of GluR3 or GluR4 mediate the rapid EPSC kinetics; flop-type receptors demonstrate four times faster desensitizing kinetics than flip-type receptors (Mosbacher et al. 1994). High content of AMPA receptors of GluR3 flop and GluR4 flop was reported in avian auditory neurons (Ravindranathan et al. 2000).

During development of NL neurons, the time course of EPSC was accelerated around the day of hatching (Kuba et al. 2002a). This implied that alterations occurred in the composition of postsynaptic receptor subunits (Brenowitz and Trussell 2001). Moreover, in the NM-NL synapses of older animals, the action potential followed more stably the train of high-frequency stimulation applied to projection fibers from NM (Kuba et al. 2002a). All these changes of synaptic transmission and membrane excitability during development improved the high-fidelity transmission or faithful processing of temporal information in the postsynaptic neurons.

2.3 Nucleus Laminaris (NL) and ITD Tuning

Nucleus laminaris (NL) encodes interaural time difference (ITD). ITD is the difference of sound arrival time between two ears and is a critical cue for localizing the sound source in the horizontal plane (Moiseff and Konishi 1981). When the sound source is located just in front of the animal, ITD is zero, while ITD increases with lateral displacement of the sound source in the horizontal plane. In birds, the binaural sound information first converges in the NL (Fig. 2.1), where the timings of spike inputs are compared and ITDs are encoded (Carr and Konishi 1990; Overholt et al. 1992).

Physiological ITD is based on the head size of animals. However, the maximum ITD calculated from the measurement of cochlear microphonic potential in the chicken was ±180 μs for low-frequency (0.8–1 kHz) and ± 100 μs for high-frequency (2–4 kHz) sound (Hyson et al. 1994). This was considerably larger than the expectation from the head size of the chicken (75 μs, a head size of 17 mm for 8–12 days after hatch). This is because of an acoustic interference of bilateral sound waves through the interaural canal, a coupling between two middle ear cavities (Calford and Piddington 1988; Larsen et al. 2006). The interaural canal connects two middle ear cavities through the cranial bone, and sound waves could be transmitted there (see the illustration in Fig. 2.34b). Accordingly, acoustic interference occurs between sound waves of two sides. The transmission was effective for low frequency, but at high frequency, the transmission through the interaural canal was almost negligible. Accordingly, the low-frequency components of sound waves are interfered significantly. Köppl and Carr (2008) reinterpreted the experimental data by Hyson et al. (1994) and extrapolated the results of the chick to matured chicken with a head size of 25 mm and estimated the physiological maximum ITD of ±160 μs for high-frequency sounds above 1.6 kHz and ± 300 μs for low-frequency sound at 800 Hz. Both were much larger than the estimation of physiological maximum ITD from animals early in postnatal days (Hyson et al. 1994).

ITD detection in the NL is highly accurate in the barn owl. When evaluated as a time window of ITD tuning, which is the ITD corresponding to the half-maximum firing probability of NL neurons, it was 100–200 μs at the frequency of 3–7 kHz (Carr and Konishi 1990; Klump 2000). This narrow time window of ITD tuning function was still larger than the resolution of ITD that was 3–6 μs in the behaving barn owl, which corresponded to the resolution of 1–2 degrees in azimuthal angle for that animal (Knudsen et al. 1979; Moiseff and Konishi 1981; Klump 2000).

2.3.1 Delay Lines

Neurons in NL have well-developed dendritic tufts in two poles projecting to the dorsal and ventral direction (Fig. 2.9). These dendritic tufts receive innervations from bilateral nucleus magnocellularis (NM). The ventral dendritic tuft was innervated by the contralateral NM projection and the dorsal tuft by the ipsilateral NM projection (Rubel and Parks 1975). The length of dendrites has a gradient along the tonotopic axis, and the low-CF NL neurons that are located in the lateral end of the nucleus have very long dendrites, and high-CF NL neurons located in the medial end of the nucleus have short dendrites (Smith and Rubel 1979).

The axonal projection of NM neurons branches to form two collaterals shortly after leaving the soma, one of which innervates the dorsal dendrites of ipsilateral NL neurons and the other the ventral dendrites of contralateral NL neurons (Fig. 2.9). Thus, a single action potential that is generated in an NM neuron triggers two spikes at the branching point of the axon. Each spike is transmitted to a series of ipsilateral or contralateral NL neurons either through the dorsal branch or ventral branch of the NM projections, respectively. The ipsilateral projection length was almost equal in individual branches from NM to NL neurons, but the contralateral projections increased the length progressively towards neurons located lateral within the nucleus. These length differences of conduction pathway functionally formed a delay line (chicken, Young and Rubel 1983, Overholt et al. 1992, Seidl et al. 2010; barn owl, Carr and Konishi 1990). The longer the conduction pathway, it takes a longer conduction time. Accordingly, the delay line makes it possible for bilateral NM spikes to arrive at coincident timing to evoke an action potential in a specific NL neuron and is proposed to compensate for the ITDs that are created externally between two ears (Fig. 2.9). NL neurons operate as coincidence detectors (CDs) of binaural temporal signals.

The difference of conduction time between ipsilateral and contralateral axonal projections of NM neurons to NL was carefully measured in the chicken using slice preparations (Seidl et al. 2010, 2014). The conduction velocity of NM axon at the

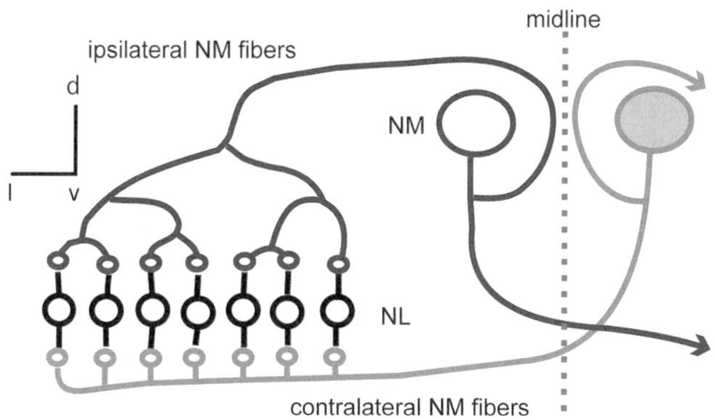

Fig. 2.9 Schema of NM projections to NL
The projection fiber of NM neuron branches shortly after leaving the soma and forms two collaterals. One branch innervates the dorsal dendrites of ipsilateral NL neurons, and, other branch, the ventral dendrites of contralateral NM neurons. Length of projection from the contralateral NM is longer toward the lateral NL neurons and forms a delay line. Length of ipsilateral NM projection is nearly the same. NL neurons operate as coincidence detectors
d dorsal, *l* lateral, *v* ventral

body temperature (40 °C) was estimated as 3.0–8.8 m/s (Overholt et al. 1992). The NM axon branched shortly after leaving the soma, and the axon for contralateral projection was longer more than 1700 μm in average than the ipsilateral one, from the branch point in the P1–P3 chicken (1–3 days after hatch, Seidl et al. 2010). Given the estimated conduction velocity at the body temperature of birds (Overholt et al. 1992), the length difference between the ipsilateral and the contralateral projections has created a too large difference of conduction time, larger than the ITD of physiological maximum range (±180 μs at P8–P12, Hyson et al. 1994). In the chicken, however, the diameter of axon of contralateral NM projection was larger (1.9 μm) than the axon diameter of ipsilateral projection (1.1 μm) and axons within the neuropil (1.4 μm) of NL (Seidl et al. 2010). The internode distance was also longer in the contralateral projection axon (about 160 μm) than the ipsilateral axon (81–93 μm) or the axons within neuropil of NL (55–61 μm). The conduction velocity of spikes in the axon is larger when the axon diameter is large. Also, the conduction velocity is larger when the internode distance is long. Based on these data of axon diameter and length of internode distance, conduction time difference could be adjusted within the range of physiological ITDs (Seidl et al. 2010). Recent report by Seidl et al. (2014) demonstrated that the conduction time was balanced between the contralateral axon and ipsilateral axon (E21 before hatching). They measured the conduction velocity of the ipsilateral axon (1.59 m/s) and the contralateral axon (3.69 m/s) at room temperature. From Q10 (1.53 in average; ranging 1.77–1.89) of the conduction velocity, they estimated that the conduction velocity at body temperature was 3.36 m/s for the ipsilateral axon and 8.02 m/s for the contralateral axon by extrapolation. The estimated conduction time from NM to ipsilateral NL was 457 ± 98 μs and that for contralateral NM was 399 ± 64 μs. Therefore, the conduction time could be balanced between two NM branches (Seidl et al. 2014). Accordingly, the compensation of ITD was not made by the axon length difference alone in the sound localization circuit. Variation in axon diameters and internode distances has provided further means to adjust the axon length disparity and played major roles to create the internal delay that compensates for the external arrival time difference of sound between two ears (ITD). An NL neuron generates action potentials with a maximal probability when bilateral inputs arrived in coincidence and encodes the ITD as location of the neuron within the nucleus where the firing probability was highest. ITD compensating mechanism of the delay line was first proposed in the MSO of mammals (medial superior olive) by Jeffress (1948).

It should be noted here that the ITD tuning curves measured in NL in vivo had many peaks and troughs that occur periodically. The period depends on neuron's characteristic frequency (CF, see Fig. 2.25). It is also noted that best frequency (BF) is frequently used in place of CF, but these two frequencies are defined in slightly a different manner; BF of neurons is defined as the sound frequency that generated spikes at the highest rate, while CF is a sound frequency at which auditory neurons have the lowest threshold for spike generation.

2.3.2 Coincidence Detection

When a pair of electrical stimuli was applied in a short time interval (Δt) in slice preparations to two projection fibers from NMs, temporally summated EPSPs reached a threshold, and the NL neuron generated an action potential (Fig. 2.10a). Neurons in high-CF NL region receive signals in high frequency and usually have a short membrane time constant. Thus, it is expected that these neurons are made suitable to integrate EPSPs when they arrive in short intervals and generate spikes. In contrast the low-CF neurons receive EPSP inputs in longer intervals. CF dependence of coincidence detection of NL neurons was evaluated in three CF regions: the high CF (2.5 kHz and higher CF, located at the rostro-medial region of NL), middle CF (1–2.5 kHz), and low CF (0.4–1 kHz, located at the caudo-lateral region of NL) (Rubel and Parks 1975, see Fig. 2.21d). It was a bit surprise to find that the time window for coincidence detection was not narrowest in the high-CF neurons

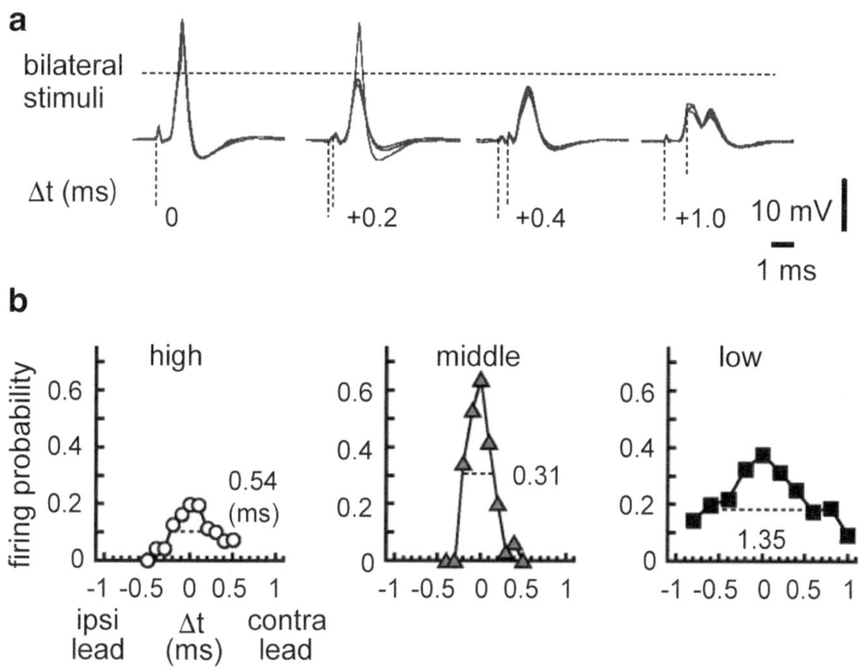

Fig. 2.10 Coincidence detection of NL neurons in slices
(**a**) bilateral NM projections to NL neurons were electrically stimulated with a time difference (Δt) between two stimuli. Spikes were generated when two stimuli were applied at close intervals. Stimulus time interval was systematically increased from the left to the right. Spike firing probability was decreased with Δt. Timing of two stimuli is indicated by vertical broken lines. (**b**) firing probability was plotted against time intervals for NL neurons of high, middle, and low-CF regions From Kuba et al. (2005)

but in the middle-CF neurons. The time window was 0.31 ms in the middle-CF neurons (when measured at the 50% of maximum firing rate), 0.54 ms in the high-CF neurons, and 1.35 ms in the low-CF neurons in slices of chicken NL after the hatch (Fig. 2.10b) (Kuba et al. 2005). A similar CF-dependent property of ITD tuning has been observed in vivo in the barn owl NL (Carr and Konishi 1990). The time window of ITD tuning curves in the barn owl was most narrow in the middle-frequency region (3–5 kHz), then in the high-frequency region (6–8 kHz), and most broad in the low (1–2 kHz) frequency region, although the absolute audible frequency range was higher than the range in the chicken (Klump 2000). Moreover, the absolute time window of CD in each frequency region was narrower in the barn owl NL than in the chicken. The time window of the barn owl NL measured in vivo was approximately one third of the time window of the chicken NL in slices.

2.3.3 Development of Animal Improves Coincidence Detection in NL Neurons

Before going into the mechanisms that determine the sharpness of CD in NL neurons, I will describe the improvement of CD during development. Figure 2.11A compared CD between the chick (P3, 3 days after hatch, Fig. 2.11Aa) and the embryo (E17, embryonic days 17, a few days before hatch, Fig. 2.11Ab), in responses to trains of four stimuli applied bilaterally to projection fibers from NM to NL in slice preparations. Bilateral stimuli were applied with a small time difference (Δt). Spikes were sharp in the chick and followed all four stimuli during the train when two stimuli were applied in coincidence. However, in the embryo spikes were broad and were generated only for the first couple of stimuli (Kuba et al. 2002a). With a small time difference between bilateral stimuli, spikes were generated only at the first stimulus pair or totally failed during the train in slices after hatch (Fig. 2.11Aa). Thus, the probability of spike generation was reduced drastic with a time separation of two stimuli. However, spikes remained firing in the embryo (Fig. 2.11Ab). Accordingly, the time window was narrower in the chick (1.4 ms) than in the embryo (3.9 ms) (Fig. 2.11B). These spikes were measured from neurons in the middle-CF region of NL, and experiments were conducted at room temperature (20–25 °C).

The membrane time constant determines the speed of electrical response of neurons. EPSP was faster in the chick (P3) and was followed by small after-hyperpolarization (Fig. 2.11Ca), which was not observed in the embryonic NL neurons (E17, Fig. 2.11Cb). The membrane conductance was increased nearly five times after the hatch and reduced the membrane time constant from 18.4 ms of the embryo to 3.2 ms, while the membrane capacitance remained nearly constant (Kuba et al. 2002a). The increased membrane conductance reflected the increased level of membrane excitability.

The membrane excitability was enhanced after hatch (Kuba et al. 2002a). Threshold of action potentials became lower, and a large after-hyperpolarization

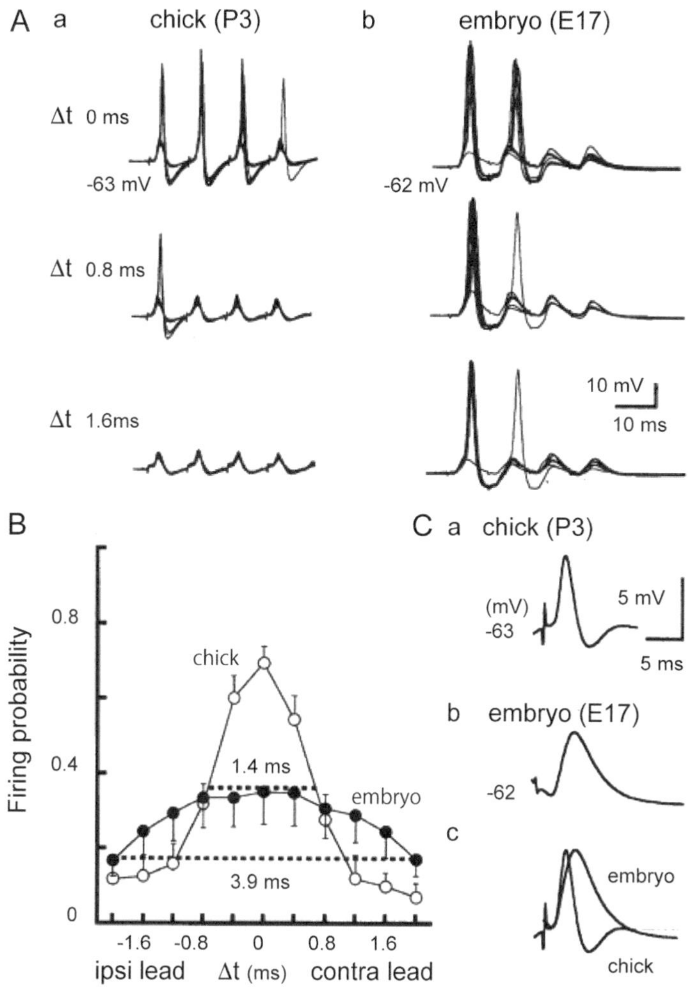

Fig. 2.11 Effects of development on coincidence detection of NL neurons in slices
(A) four pairs of electrical stimuli were applied to bilateral NM projections with time difference (Δt), and the coincidence detection was measured in P3 chick (Aa) and in embryo (E17) (Ab). (B) coincidence detection in the chick (P3, open circles) and the embryo (E17, filled circles). (C) time course of EPSPs are compared between the chick (P3) and the embryo (E17). In (Cc) amplitude of EPSP was scaled to compare the time course
From Kuba et al. (Kuba et al. 2002a, 2002b)

followed spikes in slices from the chick. The timing of spike firing became less fluctuating to the current injection (compare spikes in Fig. 2.11Aa and Ab). Conductance was increased after hatch four times for the dendrotoxin I (DTX-I)-sensitive low-voltage-activated K^+ (KLVA) channels and six times for hyperpolarization-activated cyclic nucleotide-gated cation channels (HCN), both of which accelerated the EPSP time course significantly.

It should be noted that EPSP time course was affected greatly by the experimental temperature, and the half amplitude width was accelerated 2.5 times between 20 °C (room temperature) and 40 °C (body temperature of birds). The time window for the coincidence detection was also affected by the temperature and was narrowed from 1.4 ms at 20–25 °C to 0.3 ms at 40 °C, in the middle-frequency NL neurons (Kuba et al. 2003).

2.4 Synchronization of Transmitter Release

Time course of EPSP was affected significantly by the synchronization of neurotransmitter release. Postnatal development improved the synchronization, as it was demonstrated in the synapse of rat MNTB, the medial nucleus of the trapezoid body (Chuhma and Ohmori 1998). The pattern of synaptic transmission was compared in slice preparations of rat brainstem including MNTB in postnatal days 4–14 (P4–P14). In P4–P6 animals (Fig. 2.12a), EPSCs occurred in marked asynchrony and were followed by many spontaneous mEPSC (miniature EPSC) events, when the electrical stimulation was applied to the projection fibers from contralateral AVCN (anteroventral cochlear nucleus). With development of animals, the evoked synaptic events became more synchronized; spontaneous activity disappeared, and a large single EPSC was generated after the stimulus (P9, Fig. 2.12b). The EPSC rise time became faster. However, the amplitude of mEPSCs did not change throughout the postnatal days investigated (−30 ± 0.3 pA at −70 mV). Surprisingly, the time course of EPSC recorded at P9 was superimposable with the average time course of mEPSCs recorded at P4 after the size scaling (Fig. 2.12c, d) (Chuhma and Ohmori 1998). This indicates that the synchronization of transmitter release was highly enhanced in the mature terminal. Many transmitter vesicles were released in synchrony as a single event. The projection from AVCN forms a large calyx-shaped presynaptic terminal on the MNTB neuron. The terminal was so large and accessible with a patch electrode (Fig. 2.13). A direct patch recording of presynaptic terminal demonstrated the increase of Ca^{2+} current approximately twice during these days of postnatal development (between P5–P6 and P10–P11). Accordingly, the highly phasic transmission in the mature synapse was likely achieved through the increase of Ca^{2+} current and also the Ca^{2+} sensitivity of transmitter release mechanisms in the presynaptic terminal (Chuhma and Ohmori 1998).

Ca^{2+} channel subtypes have regulatory roles in the synaptic transmission. Subtypes are classified by the sensitivity to pharmacological agents: L-type, N-type, P/Q-type, T-type, and R-type. L-type Ca^{2+} channels generally contribute for the contraction of skeletal muscle, N-type and P/Q-type Ca^{2+} channels for the synaptic transmission, and R-type and T-type Ca^{2+} channels for the neural spike activity (Schroeder et al. 2013). There are specific blockers for each subtype of Ca^{2+} channel. L-type is blocked by DHP (dihydropyridine), N-type ω-conotoxin (CgTX), and

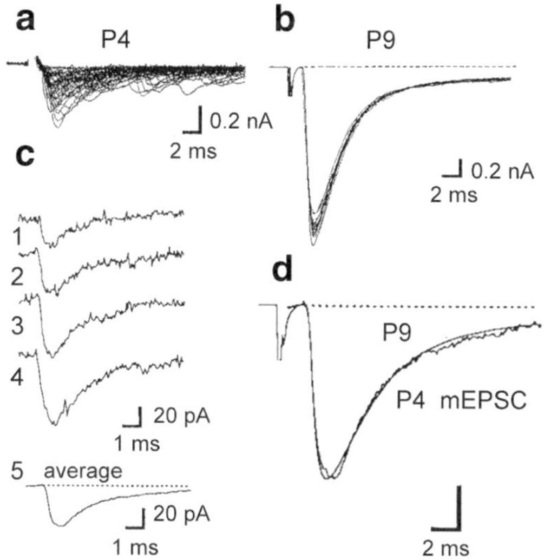

Fig. 2.12 Development of synchronization of transmitter release in MNTB synapse of the rat Slices including MNTB were made from P4 and P9 rat. The pattern of synaptic transmission was compared in NMTB principal neuron. (**a**) evoked EPSCs of P4 synapse. Electrical stimulus on the presynaptic fiber generated a burst of small EPSCs. (**b**) in P9 synapse, a single large EPSC was evoked. **a** and **b** were recorded at −70 mV. (**c**) 4 traces of representative miniature EPSCs (**c**1–4) and the average of 20 traces (**c**5). (**d**) scaled and superimposed traces of the averaged miniature EPSC (c5) and the EPSC (**b**)
From Chuhma and Ohmori (1998)

P/Q-type ω-agatoxin (Aga), but specific blocker is not found for the T-type Ca^{2+} channels. After P7 the ω-CgTX-sensitive EPSC fraction diminished and disappeared eventually after P10 from MNTB synapse, while ω-Aga-IV A-sensitive EPSC fraction increased. Presynaptic Ca^{2+} channel subtype that triggered transmitter release was apparently switched from N-type (ω-CgTX sensitive) to P-type/Q-type (ω-Aga-IV A sensitive) during the early postnatal period (Iwasaki and Takahashi 1998; Midorikawa et al. 2014), which may affect the Ca^{2+} sensitivity of synaptic transmission and might have contributed to the high-fidelity transmission in the matured synapse of auditory brainstem. Iwasaki and Takahashi (2001) further investigated and estimated the size (N) of the readily releasable pool (RRP) of synaptic vesicles and the release probability (p) from the cumulative amplitude histogram of EPSCs during high-frequency stimulation. From P7 to P14, N increased twice, whereas p decreased by a similar extent. They concluded that the developmental decrease in the release probability will establish a stable synapse transmission and transmitters were released only from a small fraction of RRP during the high-frequency transmission.

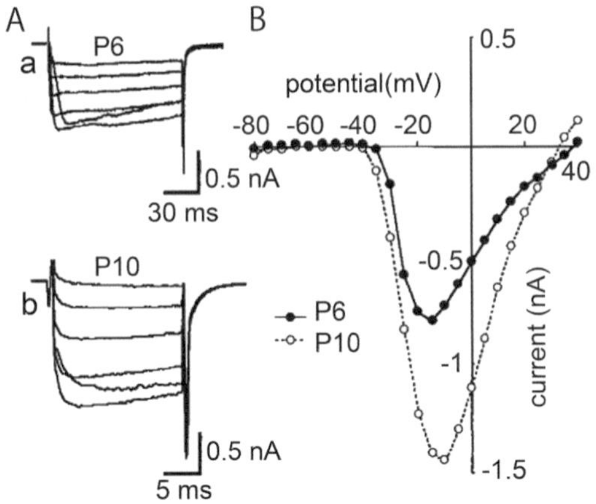

Fig. 2.13 Ca^{2+} currents recorded from the presynaptic terminal on NMTB principal neuron in slices of rat (A) Ca^{2+} currents in presynaptic terminals were increased twice in amplitude between P6 (Aa) and P10 (Ab). (B) I–V relationships of Ca^{2+} currents (P6, filled circles, and, P10, open circles). From Chuhma and Ohmori (1998)

2.5 Sharpness of EPSP and CD Has a Correlate with Expression Level of Kv1.2

When the time course of EPSCs and EPSPs were compared across CF regions of NL, EPSCs were prolonged toward the low-CF region; in contrast the time course of EPSPs was the fastest in the middle-CF region (Fig. 2.14) (Kuba et al. 2005). These were experiments conducted at chicken's body temperature 40 °C in slices of NL after hatch. Remarkably, in the middle-CF region, EPSP took a shorter time course than EPSC when the membrane potential was made a little positive (Figs. 2.14C and 2.15) (Kuba et al. 2003). EPSP is the outcome of charging the membrane capacitance by EPSC. Therefore, EPSP naturally takes a time course delayed to the time course of EPSC. However, here in the middle-CF NL neurons, the falling phase of EPSP was accelerated through the activation of low-voltage-activated K^+ (KLVA) channel, Kv1.2. Dendrotoxin-I prolonged the falling phase of the EPSPs (Fig. 2.15c).

The coincidence detection (CD) was compared in three CF regions of NL slices after hatch at the body temperature of the chick (40 °C) (Fig. 2.16). The time window of coincidence detection was narrowest in the neurons of middle-CF region, followed by the high-CF region, and was broadest in the low-CF region (Fig. 2.16b). The slope of coincidence function was another attribute of CD. The slope was maximum near the 50% of firing probability and was steepest in the middle-CF region (Fig. 2.16c). Considering that EPSP time course was narrowest in the middle-frequency NL neurons, there should be a correlation between the sharpness of CD and the time course of EPSPs. I will come back to this point in the following section, and a limiting acuity of CD will be discussed there for the chick (Fig. 2.18).

Neurons in NL express Kv1.2, and the level of expression was different depending on CF regions (Fig. 2.17) (Kuba et al. 2005). The expression level was highest in the middle-CF region, followed by the high-CF region, and was lowest in the low-CF region (Fig. 2.17e). This expression pattern of Kv1.2 channel was reciprocal to the pattern of time window for coincidence detection and in parallel with the

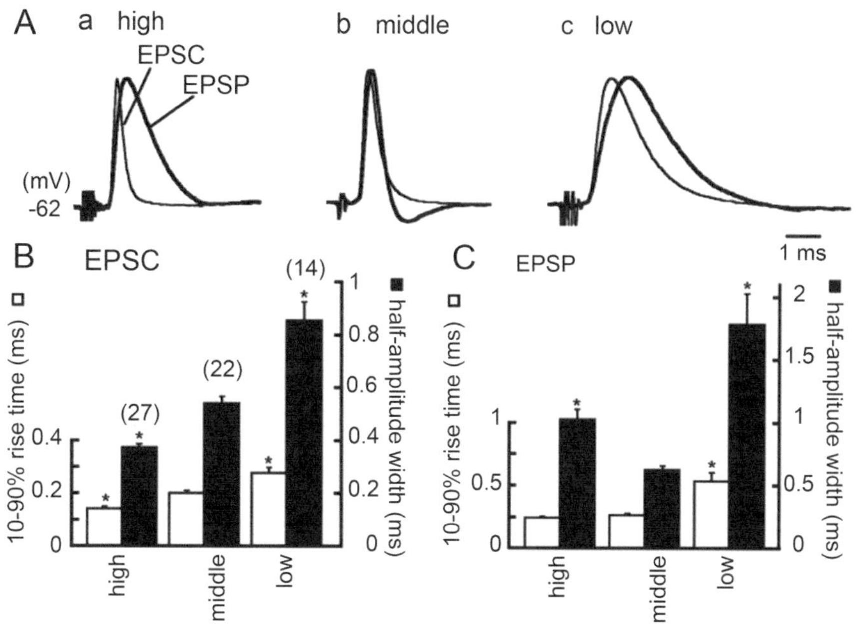

Fig. 2.14 Time course of EPSC and EPSP in NM-NL synapse
(A) EPSC (thinner lines) and EPSP (thicker lines) measured in NL neurons of three CF regions. High (Aa), middle (Ab), and low-CF NL neurons (Ac). Time course of EPSP was elongated in higher CF regions. The half amplitude width of EPSP was the shortest in middle-CF NL neurons, then the high-CF, and the widest in the low-CF neurons. (B, C) rise time (open bars) and half amplitude width (filled bars) of EPSC (B) were increased from high to low-frequency neurons. EPSP rise time was increased with the decrease of CF, but the half amplitude width of EPSP was the shortest in the middle-CF neurons (C)
From Kuba et al. (2005)

pattern of slope of coincidence detection function (Fig. 2.16b, c). Therefore, the expression level of Kv1.2 is responsible for the accuracy of coincidence detection. The CF region dependence of coincidence detection in slices has behavioral correlate with the sound localization in various animals. Most animals have the highest accuracy of sound source localization in the middle-CF region within the acoustic frequency range specific to that species (Klump 2000).

2.6 Estimation of the Limiting Acuity of Coincidence Detection

We estimated the limiting acuity of CD from the data in middle-frequency NL neurons in slices of the chick recorded at 40 °C (Kuba et al. 2003). There was a strong positive correlation between the time window of CD and EPSP time course that was measured as the half amplitude width (Fig. 2.18). Based on this correlation, we

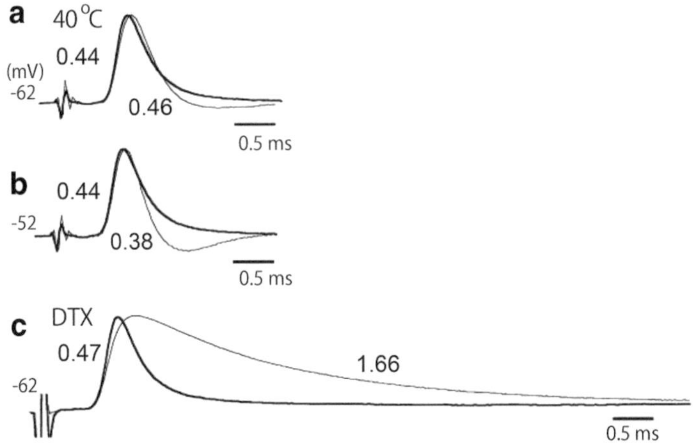

Fig. 2.15 Activation of low-voltage-activated K^+ channels accelerated the time course of EPSP (**a** and **b**) small positive shift of the holding potential (indicated on the left of traces) accelerated the falling phase of EPSP (thinner traces), but that of EPSC (thick traces) was not affected. Half amptlitude widths (ms) are indicated on the left for EPSC and on the right for EPSP. (**c**) Application of DTX (dendrotoxin, 40 nM) prolonged the falling phase of EPSP. DTX blocks the low-voltage-activated K^+ channels. Recordings were made from slice preparations at avian body temperature 40 °C
From Kuba et al. (2003)

estimated the limiting acuity of CD as 160 μs. Several observations lead to this estimation: (1) the time course of EPSP was nearly the same as that of EPSC, particularly in the middle-CF NL neurons (Fig. 2.14Ab). This was because Kv1.2 channels were activated with a small membrane depolarization and accelerated the falling phase of EPSP. Then, EPSP time course became equal to or faster than that of EPSC (Fig. 2.15). (2) With development, the time course of EPSC was accelerated (half amplitude width was three times shorter in the chick than in the embryo) (Fig. 2.11C), while the time course and amplitude of mEPSC were not different between embryo and after hatch (Kuba et al. 2002a). (3) In the rat MNTB, EPSC was also accelerated during development, while the time course of mEPSC was not different between P4 and P14 (Chuhma and Ohmori 1998). Moreover, after maturation (P9), the time course of EPSC was nearly overlapped with that of mEPSC after size scaling (Fig. 2.12d). This is because of the improved synchronization of neurotransmitter release (Chuhma and Ohmori 1998). Based on these observations, we assumed that the time course of mEPSC could be the limiting speed of EPSP in the chick NL. The median value of the distribution of mEPSC half amplitude width was 250 μs (inset trace of Fig. 2.18). The corresponding time window of CD was, thus, estimated as 160 μs from the correlation between EPSP and CD (Fig. 2.18). The time window for CD in the barn owl was about 100 μs in the middle to high-CF NL units in vivo (Carr and Konishi 1990). The limiting accuracy of CD of the chick NL

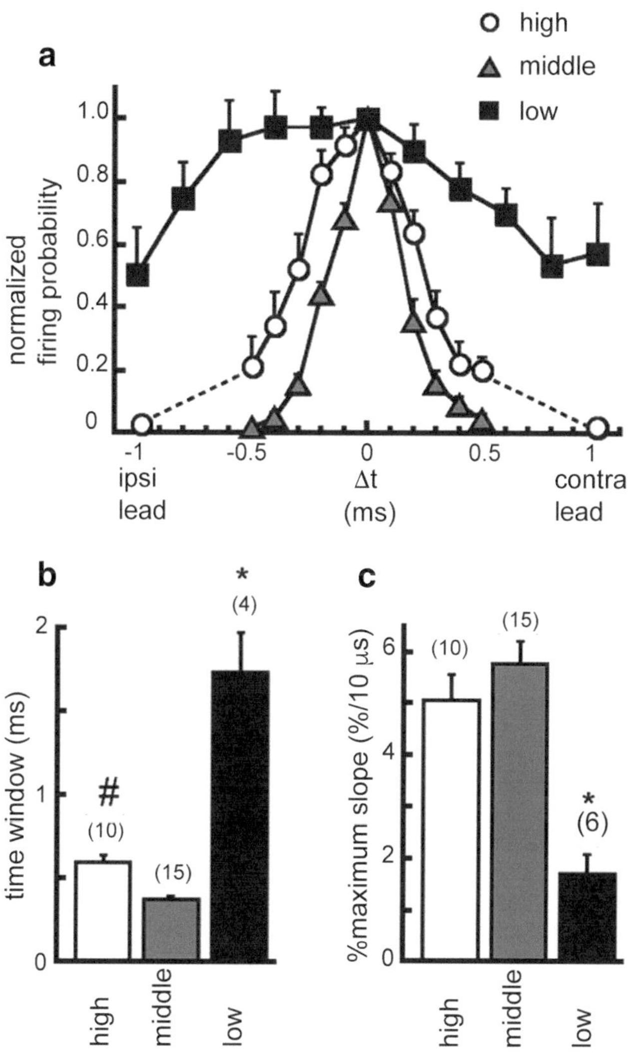

Fig. 2.16 CF region-dependent coincidence detection in NL slice preparation
(**a**) normalized plot of firing probability against the time interval (Δt) between bilateral stimulus. The time window of coincidence detection is defined as Δt corresponding to the 50% of firing rate. (**b**) time window of coincidence detection is narrowest in neurons of middle-CF, then in high-CF, and broadest in low-CF neurons. (**c**) % change of firing rate by 10 μs Δt in the coincidence detection function of three CF regions. The change was the largest in middle-CF neurons, then in high-CF, and the smallest in low-CF neurons. # $p < 0.05$, ∗ $p < 0.01$ against middle CF
From Kuba et al. (2005)

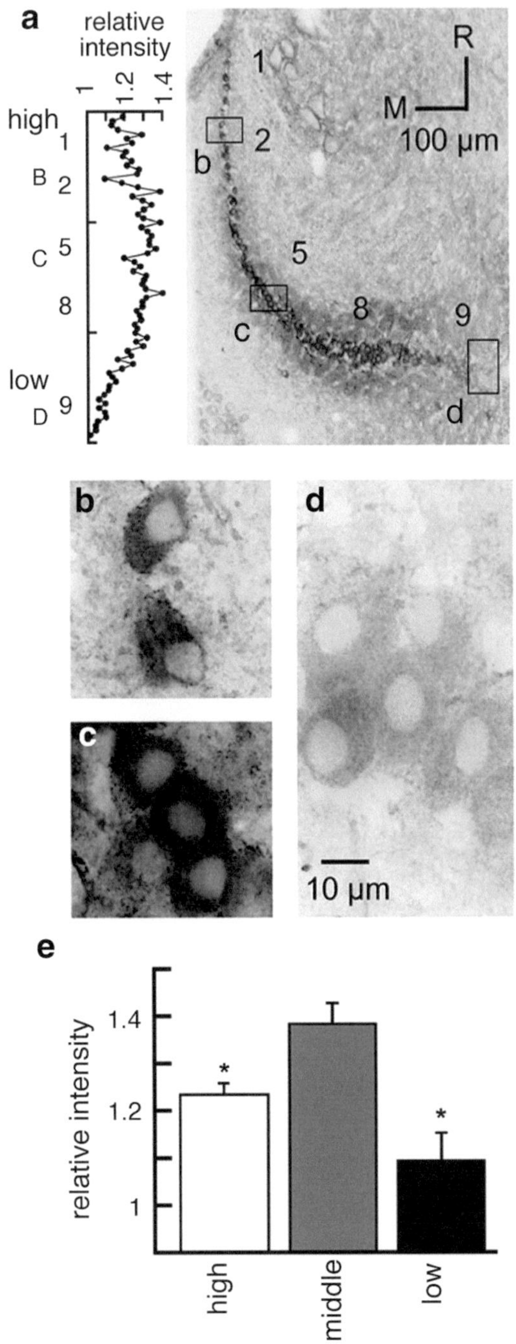

Fig. 2.17 Distribution of Kv1.2 channels along the tonotopic axis in NL (**a**) transverse slice stained with Kv1.2 antibody across high to low-CF NL regions. The left panel in **a** shows the relative intensity of immuno-labeling. Boxes show the regions of NL magnified in (**b** to **d**). (**e**) relative fluorescence intensity of the antibody expression in three CF regions. $*p < 0.01$ against middle CF
From Kuba et al. (2005)

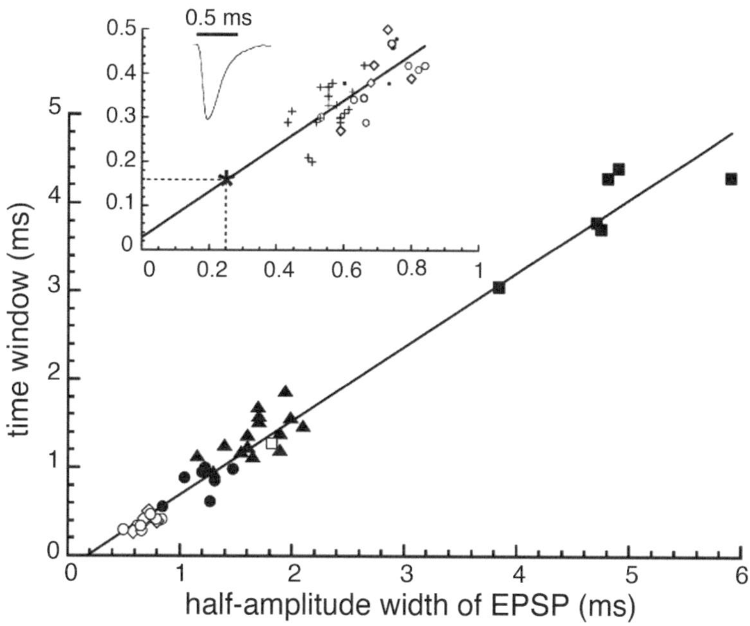

Fig. 2.18 Estimation of the limiting time window for coincidence detection
Time windows for the coincidence detection are plotted against the half amplitude width of EPSP measured in corresponding neurons. Linear fit was made to all the measurements conducted under various recording conditions, including the different ages and temperature. ■, embryo 20–25 °C (n = 6); □, embryo 40 °C (n = 1); ▲, chick 20–25 °C (n = 15); ●, chick 30 °C (n = 7); ○, chick 40 °C (n = 8); ◇, chick 40 °C using potassium gluconate-based internal solution (n = 5). Inset, plots from whole-cell (n = 13) and cell-attached experiments (n = 23), both performed at 40 °C in the chick. For cell- attached experiments: •, in the presence of bicuculline (n = 4), and +, in the absence of bicuculline in the bathing medium (n = 19). Trace of mEPSC corresponds to the median value of mEPSC distribution. Its half amplitude width is 0.25 ms. The continuous line is the least square fit to the total points at 40 °C measurements. ★, time window for coincidence detection 0.16 ms is expected from the fitted line and the time course of the mEPSC
From Kuba et al. (2003)

neuron is still broader than that of barn owl; however, the difference is small. This is just for an imagination, but it will be encouraging even for us to think that chick NL neurons might have had nearly a comparable accuracy in the sound source localization as do neurons in the barn owl NL. Barn owl is so called the superstar for the sound source localization. One of the essential differences between our slice experiments and in vivo experiments in the barn owl is the absence of effects from inhibitory controls on the CD as it will be discussed later (Fujita and Konishi 1991; Peña et al. 1996; Funabiki et al. 1998; Nishino et al. 2008; Yamada et al. 2013).

2.7 Comparison of Sensitivity of CD of the Chick with Sensitivity of ITD Tuning in Other Species

We have discussed the mechanisms that made the time window of CD narrow; however, the narrowness of time window should not be the only strategy to make detection of ITD accurate. A recent study in MSO neurons of gerbil showed that glycinergic IPSP shifted the ITD tuning function toward the contralateral leading direction by approximately 100–300 μs. Consequently, the peak of ITD tuning function was displaced at the ITD outside of the physiological range of ITD (±110–135 μs). The physiological range of ITD is determined by the head width (about 17 mm, gerbil) and the head-related transfer function (Maki and Furukawa 2005).

The peak of ITD function in MSO was located outside the physiological range of animal's ITD in many reports (Goldberg and Brown 1969; Yin and Chan 1990; Fitzpatrick et al. 1997; Brand et al. 2002). It was proposed that the slope of ITD function, rather than the location of peak or the narrowness of ITD tuning curve, was important in encoding ITDs in mammals (slope coding) (McAlpine et al. 2001; Brand et al. 2002). The possibility of slope coding was discussed also in the chicken (Joseph and Hyson 1993). As shown in Fig. 2.16c we measured the maximum slope of firing probability from the coincidence detection function (Kuba et al. 2005). The maximum slope was calculated as the change of firing probability in percentage over 10 μs of ITD or Δt (the time difference between stimuli applied bilaterally). The maximum slope was 5.8%/10 μs-Δt at 40 °C in the chick NL slice experiments (Fig. 2.16c). This value was close to those reported in other animals in vivo (5–6%/10 μs-ITD in budgerigar NL, from Fig. 6.10 in Klump 2000; 1–3%/10 μs-ITD in gerbil MSO, Brand et al. 2002). In the barn owl, the maximum slope recorded in vivo in the middle- to high-CF NL neurons was extremely steep (9–12%/10 μs-ITD, Carr and Konishi 1990). The barn owl could even localize the sound source with errors as small as 2 degrees in azimuth when the target sound was located frontally (Knudsen et al. 1979). This corresponded to the detection of about 6 μs of ongoing time disparity in space sensitive IC units (Moiseff and Konishi 1981). Accordingly, a firing rate difference of about 10% seems the basis for auditory neural system to recognize the difference.

2.8 Other Factors That Affect CD in NL

Sharp CD in NL can be attributed to the activation of low voltage-gated K^+ channels. Moreover, the CD in NL neurons are modulated by various other factors, such as the activation of HCN channels to affect the resting potential (Yamada et al. 2005), small size of EPSPs by desensitization of glutamate receptors or activation of metabotropic glutamate receptors (Kuba et al. 2002b; Okuda et al. 2013), localization of axon initial segment and distribution of Na^+ channels (Kuba et al. 2006), GABA receptors, and intracellular Cl^- concentration (Funabiki et al. 1998; Brückner and Hyson 1998). Also, activation of GABAergic neurons in networks increases the tolerance of ITD tuning to sound level (Nishino et al. 2008; Yamada et al. 2013). As such ITD tuning or CD in NL are modulated by various factors.

2.8.1 HCN Channels

Activation of HCN channel affects the coincidence detection through effects on the resting conductance of neurons (Kuba et al. 2003). Current through the HCN channel (Ih) is slowly activated during membrane hyperpolarization. The current is carried both by K^+ and Na^+ (Pape 1996). Four subtypes were described, and two of them existed in the brainstem auditory neuron: HCN1 and HCN2. Because the channel is activated by membrane hyperpolarization and the gating is sensitive to the concentration of cyclic nucleotide, the channels are termed hyperpolarization-activated and cyclic nucleotide-gated cation (HCN) channels (Ludwig et al. 1998, 1999; Santoro et al. 1998; Ishii et al. 1999; Seifert et al. 1999).

Distribution of HCN channels in NL was demonstrated by in situ hybridization technique (Yamada et al. 2005). HCN1 and HCN2 existed but not HCN3 nor HCN4. Expression level of HCN1 was low in the high-CF region but increased toward the low-CF region. In contrast, HCN2 expression was relatively constant throughout the nucleus (Fig. 2.19A).

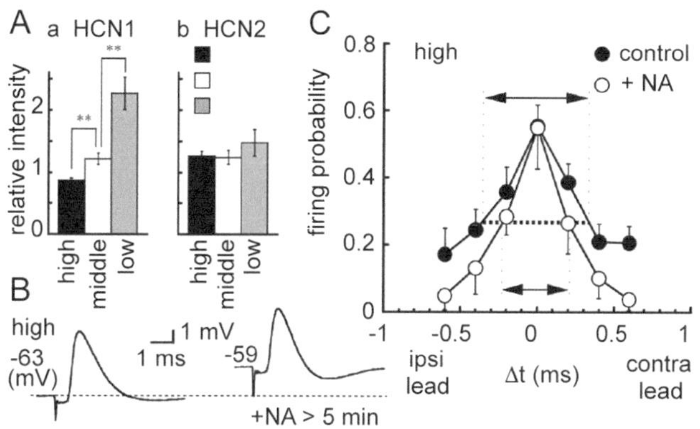

Fig. 2.19 Distribution of HCN1 and HCN2 channels in NL
(A) density of immunohistochemical labeling of HCN channel expression. (Aa), the expression of HCN1 channel was densest in the low-CF region, then in middle, and thinnest in high-CF region. (Ab), HCN2 channel expression was not different in these CF regions. (B) incubation with noradrenaline (50 μM, NA) more than 5 min depolarized the resting potential of high-CF neurons in slice preparation. (C) coincidence detection curves in high-CF neurons in slice preparations. Measurements were made in ACSF (filled circles, control) and after incubation in NA-enriched medium more than 5 min (open circles). Time window for CD is 0.7 ms in the control and 0.43 ms in NA
From Yamada et al. (2005)

HCN channel subtypes are different in sensitivity to cAMP and gating property. HCN1 channels are activated faster and at more positive membrane potentials than are HCN2 channels (Santoro and Tibbs 1999). HCN2 channels are more sensitive to the intracellular cAMP than HCN1 channels. The activation kinetics of HCN2 channels were accelerated by the increase of intracellular cAMP, and the voltage dependence was shifted to the positive direction.

Concentration of cAMP can be regulated through G-protein-coupled membrane receptor mechanisms (Gilman 1987). Monoamine and acetylcholine applied extracellularly modulated Ih by the cyclic nucleotide-dependent processes (DiFrancesco et al. 1986; DiFrancesco and Tromba 1988; Bobker and Williams 1989). In neurons of MNTB of the rat, noradrenaline shifted the voltage dependence of Ih gating to a more depolarized potential (Banks et al. 1993).

In the chicken NL slices of high-CF region, bath application of noradrenaline (NA, 50 μM) accelerated the activation kinetics of Ih. The resting membrane potential was depolarized by 5 mV (Fig. 2.19B). The EPSP time course was accelerated, and the EPSP amplitude was reduced significantly. In contrast in neurons of the middle-CF region, none of these changes were induced by NA incubation (Yamada et al. 2005). The depolarization in the high-CF neurons activated the low-voltage-activated K^+ channels, accelerated the time course of EPSP, reduced the EPSP amplitude, and affected the precision of coincidence detection. The time window became narrower after the incubation with NA (Fig. 2.19C). Distribution of noradrenergic fibers were demonstrated in numerous regions of the brainstem including NL (Moons et al. 1995), suggesting a possibility of improving the sound source localization under some condition of sympathetic activity through enhancing Ih.

2.8.2 *Effects of Depression of Synaptic Transmission on Coincidence Detection*

ITD sensitivity could be affected by the size of EPSPs. In experiments conducted first in the NL slice preparations using extracellular recording method (Joseph and Hyson 1993), weak bilateral electrical stimulation produced a narrower time window for the coincidence detection than the case of stronger bilateral stimulation. Later intracellular recording experiments, conducted also in NL slice preparations, demonstrated that the faster kinetics and smaller size EPSPs improved the coincidence detection (Kuba et al. 2002b). During trains of stimulation, size of EPSPs was depressed. This EPSP depression was attributed primarily to the reduction of transmitter release from presynaptic terminal, because the level of depression was relieved after reducing the transmitter release by Cd^{2+} (2.5 μM), but was not affected by reduced desensitization of AMPA receptors on the postsynaptic neurons with cyclothiazide (20 μM). Cd^{2+} is a general blocker of Ca^{2+} channels.

We evaluated the accuracy of coincidence detection in the NL slice preparation, by applying bilateral trains of 4–10 stimuli at 100 Hz (Kuba et al. 2002b). The first

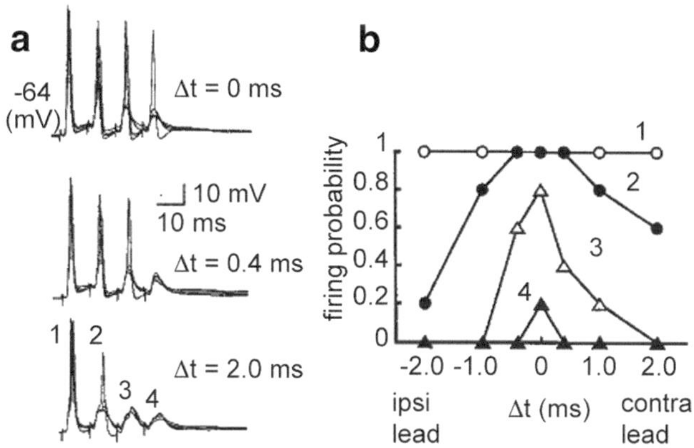

Fig. 2.20 Depression of synaptic transmission improved the coincidence detection
(**a**) train of four stimuli was applied to bilateral NM projection to NL with time interval Δt. With increase of Δt, firing probability was decreased particularly at the stimulus timing late in the train. (**b**) firing probability was plotted against Δt separately for the first to the fourth stimulus in the train. The time window for coincidence detection was narrowed toward the late stimulus. Recordings were made at room temperature (20–25 °C) from brainstem slices of chick embryo (E16–18)
From Kuba et al. (Kuba et al. 2002a, 2002b)

stimulus in the train generated spikes in 100% of trial (Fig. 2.20a). The time window for the coincidence detection was narrowed progressively during the course of stimulus train (Fig. 2.20b). This occurred in parallel with a decrease of EPSP amplitude. In addition, the reduction of EPSP amplitude by Cd^{2+}(2.5 μM) or CNQX (2 μM) similarly improved the accuracy of coincidence detection (Kuba et al. 2002b; Cook et al. 2003). These results indicate that the synaptic depression improves the coincidence detection in the chick NL neurons.

2.8.3 *Effects of Metabotropic Glutamate Receptors on Coincidence Detection*

Small subthreshold EPSPs were not able to trigger spikes, although small EPSPs improved the CD (Kuba et al. 2002b). Accordingly, it was ideal to have some mechanism to optimize the size of EPSPs during ongoing stimuli. Several different types of presynaptic receptors are known to regulate transmitter release by modulating calcium and potassium currents (Barnes-Davies and Forsythe 1995; Lachica et al. 1995; Otis and Trussell 1996; Takahashi et al. 1996, 1998; Leão and Von Gersdorff 2002). During tetanus stimulation, metabotropic glutamate receptors (mGluRs) were activated. Metabotropic GluRs are known to operate as autoreceptors at glutamatergic presynaptic terminals and to depress the release of neurotransmitter (Conn and Pin 1997). Metabotropic GluRs reduced the size of EPSPs, reduced the progress of EPSP depression, and improved the CD in slices of NL (Okuda et al. 2013).

A non-specific agonist of mGluRs ((±)-1-aminocyclopentane-trans-1, 3-dicarboxylic acid; t-ACPD) reduced the amplitude of EPSCs and diminished the progress of depression during a stimulus train (Fig. 2.21a, b), while the paired pulse ratio (PPR) and coefficient of variation of EPSC amplitudes were increased. These observations are consistent with the reduced release probability of neurotransmitters. Thus, the effects of t-ACPD were presynaptic and reduced the release of neurotransmitter from the terminal (Okuda et al. 2013).

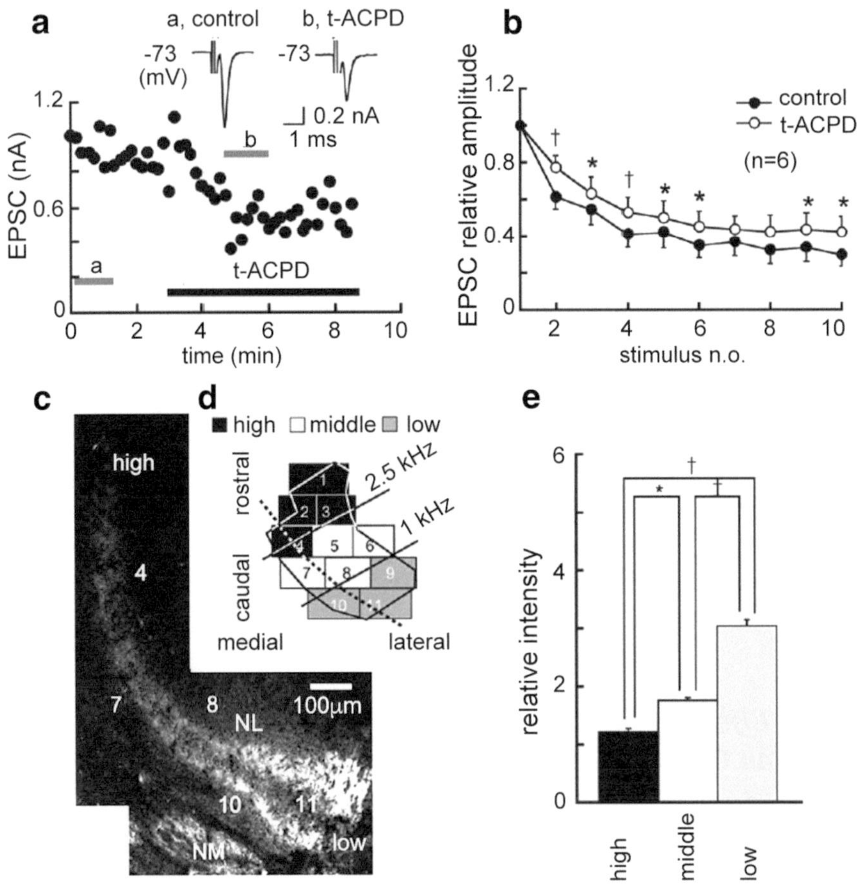

Fig. 2.21 Effects of t-ACPD on EPSCs and immunohistochemistry of mGluR
(**a**) EPSC amplitude was reduced by bath-applied t-ACPD. Inset 2 traces show the average EPSC during bars at *a* and *b*. (**b**) amplitudes of EPSCs relative to the first EPSC in the train. A train of ten stimuli was applied to NM-NL projection at 10 ms inter-pulse intervals. Depression progressed slower in t-ACPD ($n = 6$). *$p < 0.05$; †$p < 0.01$. (**c**) fluorescence image of mGluR2/3 immunohistochemistry from a slice containing three CF regions of the NL along the broken line in **d**. Numbers correspond to areas in the nucleus labeled in **d**. (**d**) a scheme to show CF regions of NL. (**e**) intensity of immuno-labeling for mGluR2/3 is plotted for three CF regions of NL. *$P < 0.05$ and †, $P < 0.01$
From Okuda et al. (2013)

Expression of group II mGluRs (mGluR2/3) was graded along the tonotopic axis in the NL and was stronger toward the low-frequency region (Fig. 2.21c, d, e). The level of expression was increased after hatch particularly in the low-frequency region of the nucleus. Both group II (DCG-IV)- and group III (L-AP4)-specific agonists reduced EPSC amplitude, and the extent of reduction was larger in the low-frequency NL region. The reduced EPSP amplitude in DCG-IV improved the CD (Okuda et al. 2013). There were no detectable effects of group I-specific agonists on EPSCs.

A specific antagonist of group II mGluRs (LY341495) increased the amplitude of both EPSCs and EPSPs and enhanced the depression during a stimulus train, indicating that mGluRs were activated constitutively in the NL. These observations indicated that mGluRs likely operated as autoreceptors and regulated EPSP size to improve the CD in the NL.

2.8.4 Na^+ Channel Distribution at the Axon Initial Segment

Axon hillock is myelinated in the high-CF neuron but not in the low-CF neurons of the chicken NL (Carr and Boudreau 1993). It was also demonstrated by immunostaining of Na^+ channels and retrograde tracer label of the NL neuron that axon initial segment (AIS) is arranged at some distance from the soma (Kuba et al. 2006). Na^+ channels were not found in the soma of NL neurons, but they were clustered in the AIS at the location separated by 20–50 μm from the soma. In low-CF neurons, Na^+ channels were clustered in a longer stretch of AIS but were located closer to the soma (Fig. 2.22c, d, e). Thus, neurons likely have different patterns of AIS and initiate spikes at more remote site as the increase of CF (Fig. 2.22).

In addition to the observation based on immunostaining, there were consistent electrophysiological evidences that indicate a distant location of spike initiation in the high to middle-CF neurons (Fig. 2.23), such as:

(1) Both antidromic and orthodromic spikes were small in high-CF and middle-CF neurons, but those spikes were large in low-CF neurons (Fig. 2.23a).
(2) Antidromic spikes were not affected by somatic subthreshold membrane depolarization in the middle to high-CF neurons. In contrast in the low-CF neurons, antidromic spikes were affected by somatic depolarization, which interrupted eventually the spike generation (Fig. 2.23b).
(3) By cell-attached patch recordings from the soma, small inward going Na^+ current was occasionally observed in the low-CF neurons, but never from the high to middle-CF neurons. Whole-cell recording demonstrated a small TTX-sensitive inward current from high to middle-CF neurons, but the inward current was five to six times larger in the low-CF neurons (Kuba et al. 2006).

Computer simulation demonstrated that the geometry of AIS was optimized in the distance from the soma and in the segment length to reduce the threshold and to stabilize the spike generation. The threshold current could be lowest when AIS was

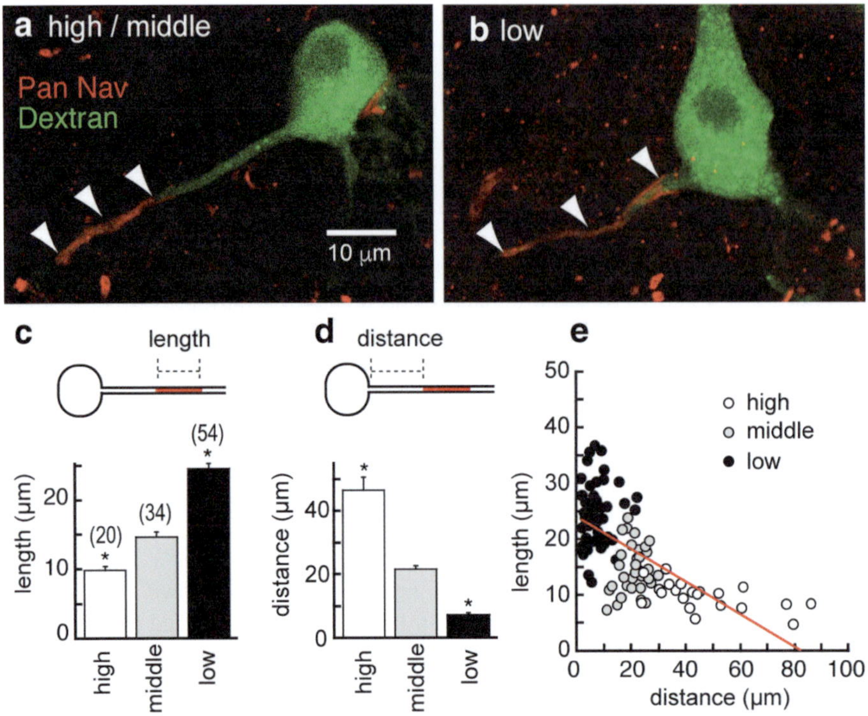

Fig. 2.22 Distribution of Na^+ channels in axon initial segment in NL
(**a**, **b**) NL neurons were retrograde labeled (green) from IC, and Na^+ channels were immuno-labeled (red). Note Na^+ channel immuno-positive regions are stretched on the axon and located in a distance from the soma in high and middle-CF neurons (**a**) and closer to the soma in low-CF neuron (**b**). Axon initial segment (AIS) is indicated by arrow heads. (**c**) the length of AIS is the longest in low-CF neurons and shortest in high-CF neurons. (**d**) AIS is located most distant from the soma in high-CF NL neurons and closest to the soma in low-CF neurons. (**e**) the length and distance of AISs from the soma are reciprocally related
From Kuba et al. (2006)

located next to the soma. However, spike initiation was likely suspended quickly by inactivation of Na^+ channels by the depolarization of somatic membrane particularly when high-frequency excitatory inputs were made. At a distance from the soma, the level of depolarization should be reduced by the cable property of the axon, which reduced the inactivation of Na^+ channels. The high-CF neurons generated spikes stably, with the lowest threshold current when the AIS (10 μm length) was distantly located (20–30 μm). However, further displacement increased the threshold current, either by the dissipation of current across the axonal membrane or the increased K^+ current. In low-CF neurons, the closer location of AIS to soma reduced the threshold current, while the accumulated Na^+ channels in the long AIS overcame the inactivation resulting from the depolarization of low-frequency excitatory inputs (as it was already discussed in low-frequency NM neurons, Fig. 2.7Bd). The level of depolarization was smaller than in the case of high-CF neurons because

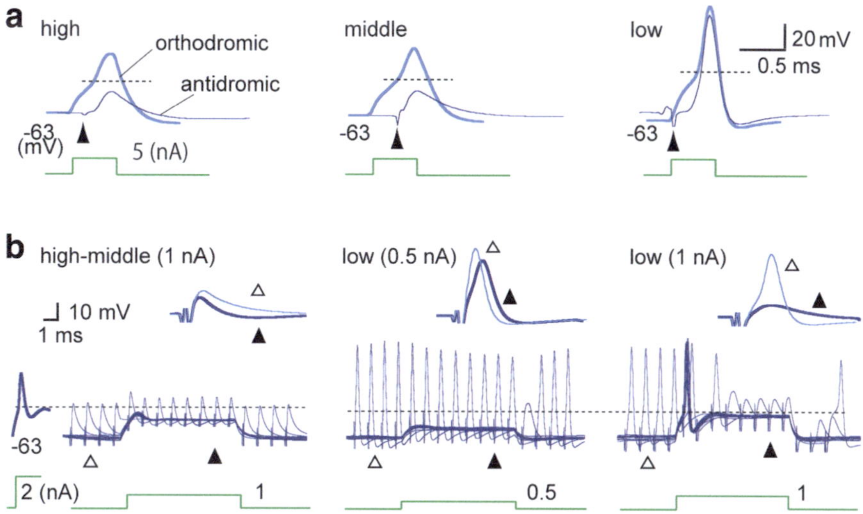

Fig. 2.23 Spike size is different between low-CF and high- or middle-CF neurons in NL (**a**) spikes were generated orthodromic by current injection to the soma (thick blue lines) or antidromic by electrical stimulation applied to the axon (thin blue lines). Antidromic stimulation was applied at the timing of arrowhead. Spike size was larger in low-CF neurons than those of high or middle-CF neurons. (**b**) effects of somatic depolarization on antidromic spike generation. Antidromic spikes (△) were small and were not affected by somatic depolarization (▲) in high or middle-CF neurons. The antidromic spike was large (△) and was affected (▲) by somatic depolarization (low, 0.5 nA) and eventually aborted by larger depolarization (▲, low 1 nA) in the low-CF neurons. These different effects of somatic depolarization may reflect closer location of spike generation site to soma in low-CF neurons than in high or middle-CF neurons
From Kuba et al. (2006)

of lower incidence of excitatory inputs. Thus, the distance and the length of AIS were arranged to increase the excitability of NL neurons, which eventually improved the ITD sensitivity in a CF-dependent manner (Fig. 2.22e).

Additionally, Na^+ channels were not detectably distributed in the cell soma of NL neurons (Fig. 2.22a, b). This absence of Na^+ channels was important to accelerate spikes of NL neurons. If spikes were invaded in soma, speed of spike generation and conduction was significantly affected by the somatic capacitance and the activation of K^+ current (Hossain et al. 2005).

Furthermore, the structure of AIS in NL neurons was modulated in the activity-dependent manner. Before hearing onset, the ankyrin G-positive AIS existed in a wide stretch of proximal axon regardless of the tuning frequency. Shortly after the hearing onset, Na^+ channels accumulated along the entire AIS. Then AIS started shortening and relocating distally in neurons with higher tuning frequencies (Kuba et al. 2014). Ablation of inner ears abolished the shortening of AIS without affecting the position of its proximal boundary. These observations indicated that both distal and proximal boundaries of AIS were rearranged during development. The rearrangement determines the cell-specific distribution of AIS and plays a critical role in fine-tuning of ITD processing (Kuba et al. 2014).

2.9 Inhibitory Synaptic Mechanisms to Improve the ITD Processing

Low-voltage-activated K^+ channels basically play critical roles to make coincidence detection sharp in the NL. Moreover, inhibitory networks play modulatory roles in the ITD tuning. GABAergic synapses modulate the coincidence detection and affect the ITD tuning in the NL in vivo in CF and sound level-dependent manner.

2.9.1 Effects of GABAergic Inhibition on the Coincidence Detection

Effects of GABA on the performance of ITD processing were demonstrated in the inferior colliculus of the barn owl (Fujita and Konishi 1991). GABAergic innervations were anatomically reported in NL from the superior olivary nucleus (SON) or possibly from the interneurons located near NL and NM (von Bartheld et al. 1989; Code et al. 1989; Lachica et al. 1994). We, therefore, tested the role of inhibitory innervation on the ITD processing first in slice preparations of NL from embryonic chick (E15–20) (Funabiki et al. 1998).

GABAergic postsynaptic potentials (PSPs) could be either hyperpolarizing or depolarizing depending on the Cl^- gradient between the intracellular and extracellular spaces. GABAergic PSPs are usually depolarizing in immature neurons and change to hyperpolarizing in mature neurons, due to the upregulation of Cl^- extrusion mechanisms with development (Cherubini et al. 1991; Shibata et al. 2004; Watanabe and Fukuda 2015). Funabiki et al. (1998) tested whether the GABAergic PSPs were depolarizing or hyperpolarizing in embryonic NL neurons. GABAergic postsynaptic currents (PSCs) were isolated by blocking glutamatergic activity by CNQX (antagonist of AMPA receptors). NMDA receptor currents were not detected in the NL neuron. Even after application of CNQX, NL neurons demonstrated frequent PSCs events, which were blocked by bicuculline (20 μM), thus GABAergic. The polarity of PSCs was inward going at the resting membrane potential (-60 ± 2 mV) immediately after the start of whole-cell recording, indicating that the reversal potential of Cl^- was more positive than the resting membrane potential and the concentration of intracellular Cl^- should be high (Funabiki et al. 1998). The polarity of PSCs was changed to outward going with time after the whole-cell recording (15–20 min), when recorded with a low Cl^- pipette medium (5 mM Cl^-, 155 mM K^+-gluconate intracellular medium, E_{Cl} -90 mV). Therefore, GABAergic inputs likely depolarize the postsynaptic NL membrane in the embryonic chick (E15–20). In recording with a high Cl^- pipette medium ($E_{Cl} - 45$ mV), extracellular application of GABA (10 μM) depolarized the membrane potential (from −67 mV to −58 mV) and reduced the input impedance and size of EPSP (EPSP was reduced

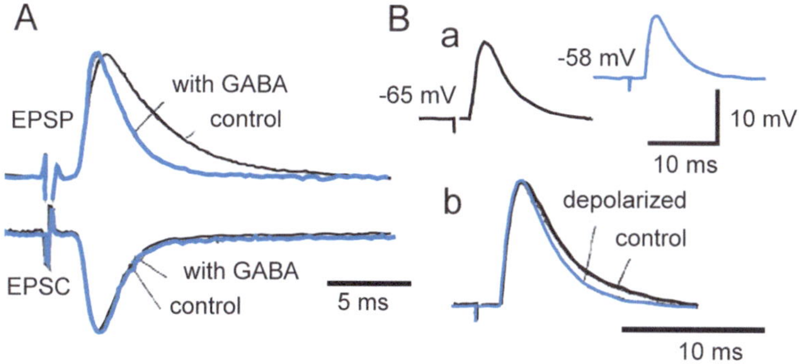

Fig. 2.24 GAGA effects on EPSPs in slices of NL
(**A**) application of GABA (10 μM) accelerated the decay of EPSP but did not affect the time course of EPSC in slices of chick embryo (E15–20). Membrane potential was held at −72 mV. Traces were scaled in amplitude to facilitate comparison of time courses. (**Ba**) membrane potential was depolarized by current injection to the level induced by GABA application. (**Bb**) depolarization slightly accelerated the falling phase of EPSP. EPSP sizes are scaled
From Funabiki et al. (1998)

to 43% of the control) (Fig. 2.24A). Moreover, the time course of EPSP was accelerated. Simple depolarization of the membrane potential by current injection reduced EPSP size (Fig. 2.24B). But the size reduction and the acceleration of time course were smaller than those induced by GABA application. Accordingly, GABA application likely accelerated the EPSP through shunting of input impedance of the neuron.

In view of the accelerating effects of GABA on the postsynaptic potential, GABA likely contributes to improve the coincidence detection in NL (Funabiki et al. 1998). We tested the coincidence detection by combining evoked EPSPs with current injections through the recording patch pipette. EPSPs were evoked by electrical stimulation applied to the ipsilateral NM projection fibers, and the contralateral inputs were emulated by injection of depolarizing current to the NL neuron. The duration of injected current pulse was set nearly equal to the mean duration of the recorded EPSP. Local GABA (10 μM) application improved the coincidence detection (Funabiki et al. 1998). GABA application depolarized the membrane from −67 mV to −61 mV, accelerated the decay time course of EPSP, and improved the coincidence detection in all neurons tested. The time window was reduced to 30% of the control. Accordingly, GABAergic synapse activity improved the coincidence detection in the embryonic NL neurons. In contrast, the membrane depolarization of 7 mV by current injection alone did not narrow the response window. This lack of effects of membrane depolarization should be due to immaturity of KLVA channels in the embryonic chick. Consistently, EPSP was monophasic in the chick embryo (E17), while after hatch (P3), EPSP was followed by after-hyperpolarization (Fig. 2.11c).

2.9.2 GABA$_B$ Receptor Activation

Depolarization of the membrane potential by GABAergic synapse was elegantly confirmed in the laboratory of Dr. Lu, by using a gramicidin-perforated patch electrode recording (Tang et al. 2009a, b). They made slice preparations from chick embryos (E19–21) and showed that the reversal potential of GABAergic response was more depolarized (−36 mV) than the spike threshold (−44 to −49 mV) or the resting potential (mean −59 mV) of NL neurons. Therefore, activation of GABAergic inputs produced a mixture of inhibitory and excitatory actions. The inhibitory action improved the acuity of temporal processing of sound signal (Funabiki et al. 1998). However, because the depolarization enhanced excitability of neurons, it might have reduced the phase-locked firing property of NL neurons. Tang et al. (2009a, b) proposed that the activation of presynaptic metabotropic glutamate receptors (mGluRs) or GABA$_B$ receptors could reduce the transmission of GABA synapse at the presynaptic terminal while leaving the postsynaptic inhibitory action intact. As expected, application of baclofen (agonist of GABA$_B$ receptors) reduced the frequency of miniature IPSCs (to 38 ± 21% of the control) without changing individual IPSC amplitudes, while the activation of mGluRs reduced the IPSC size.

To date, eight members of mGluRs are identified (see Sect. 2.8.3). They are classified into three groups based on pharmacological features and signal transduction pathways: group I, mGluR1 and mGluR5; group II, mGluR2 and mGluR3; and group III, mGluR4, mGluR6, mGluR7, and mGluR8 (Kew and Kemp 2005). Since some glutamatergic terminals and GABAergic terminals were located close each other, it was expected that GABAergic transmission can be modulated by activation of some mGluRs. As predicted t-ACPD (non-specific agonist of group I and II mGluRs) reduced the amplitude of IPSC. More specific agonists in group II (DCG-IV) and group III (L-AP4) mGluRs affected GABA release, but the one of group I (3,5-DHPG) mGluRs did not. Accordingly, GABAergic responses were reduced effectively in NL neurons by presynaptic activity of GABA$_B$ receptors or metabotropic glutamate receptors. Overall, GABA$_B$ receptors and mGluRs may regulate the basal level of GABA release from inhibitory synapse in NL.

2.9.3 GABAergic Synapse from SON Improves Low-Frequency ITD Processing

GABA terminals are concentrated around the perikaryon and proximal dendrites of NL neurons (Carr et al. 1989; Code et al. 1989; Carr and Boudreau 1993). Major source of these terminals is the ipsilateral superior olivary nucleus (SON). SON receives inputs from the nucleus angularis (NA), of which activity is proportional to the sound level. Therefore, SON activity is sound level dependent. SON makes inhibitory innervations to the ipsilateral NL, NM, NA, and contralateral SON (Fig. 2.1) (Lachica et al. 1994; Yang et al. 1999; Burger et al. 2005). This inhibition from SON in the NL was proposed to increase the threshold of neural activity and to make the ITD sensitivity tolerant to the sound level in the barn owl (Peña et al. 1996). It was further hypothesized that SON activity balances the influence of interaural level difference (ILD) in the processing of ITD (Burger et al. 2005; Dasika et al. 2005).

In the chicken of postnatal days from 3 to 29 (P3–P29), the peak-trough contrast of ITD, which is the ratio of firing rates between the peak and the trough of ITD tuning curve, was enhanced in low BF NL neurons (<1 kHz) when a loud sound was applied. However, the contrast was reduced in high-BF NL neurons (Fig. 2.25a, b) (Nishino et al. 2008). The firing rate of NL neurons should be increased with the rise of sound level, but surprisingly in low BF NL neurons, the firing rate was suppressed at the worst ITD with the increase of sound level (Fig. 2.25a). In contrast in the middle-BF and high-BF (≥1 kHz) NL neurons, the firing rate was not reduced at the worst ITD when a loud sound was applied. In these neurons, the loud sound increased the firing rate both at the best and the worst ITDs. Thus, the contrast of ITD tuning curve disappeared (68 dB, Fig. 2.25b).

The suppression of firing activity at the worst ITD was eliminated by electrolytic lesion of the SON ipsilateral to the NL recording site (Fig. 2.25c). Accordingly, the suppression at the worst ITD was managed by the inhibitory projection from SON to the low-frequency region of NL. The SON projection to NL was visualized by transfection of neurons in SON using Sindbis virus that encoded RFP (red fluorescence protein). The projection was most dense in the low-frequency region of NL, then in the middle, and the least dense in the high-frequency region (Fig. 2.26a, b, c). Moreover, these RFP-labeled terminal boutons were immunoreactive to VGAT (vesicular GABA transporter) (Fig. 2.26d, e, f). The VGAT immunoreactivity overlapped with 89% of the RFP-labeled SON terminals in low-CF, 93% in middle-CF, and 84% in high-CF region (Nishino et al. 2008). Therefore, the inhibitory projection from SON was dependent on the CF region in NL and was GABAergic.

With the SON inhibition, the best phase of unit activity occurred within a narrow phase angle of sinusoidal sound in low BF NL neurons (vector strength was close to 1.0, BF 200 Hz), when a loud sound was applied (98 dB SPL) (Fig. 2.27a control). After the lesion of SON, thus in the case without inhibition, the firing activity in NL neuron of BF 400 Hz still occurred in a good phase-locked manner when sinusoidal sound was applied (vector strength was around 0.8, 38 dB SPL, Fig. 2.27a lesion). However, the best phase was displaced with ITD, and the contrast of ITD tuning was reduced (Fig. 2.27b lesion). At much higher levels of sound stimulus, the contrast of ITD tuning totally disappeared as it was so in the case of high-BF NL units where SON was intact (Nishino et al. 2008).

The degree of SON inhibition was dependent on the sound level. When the sound level was high, the SON inhibition suppressed effectively the spike firing of NL neurons near the trough of ITD tuning curve where the firing activity was low, but did not suppress firings near the peak of ITD tuning curve where the firing activity was robust. Thus, the firing activity of NL neuron was limited within a limited phase or a limited ITD. Accordingly, the contrast of ITD tuning was enhanced significantly (Nishino et al. 2008). This inhibitory mechanism of SON and its roles in the

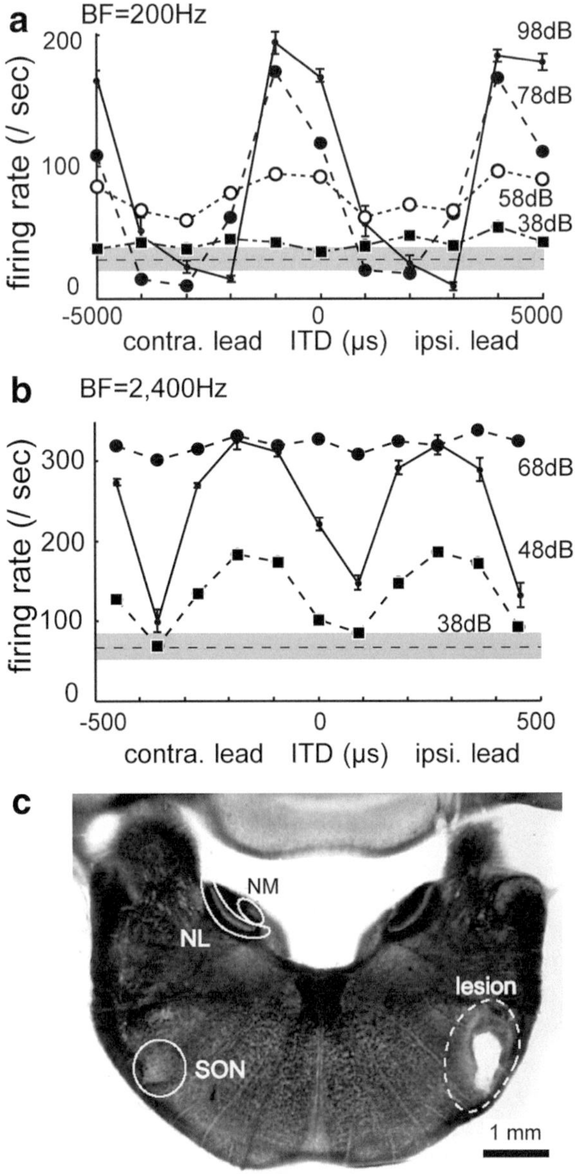

Fig. 2.25 Sound level effects on ITD tuning curves in NL of the chick in vivo
(**a**) ITD tuning curves measured in 200 Hz BF NL neurons. Binaural sound with ITD was applied at the sound levels indicated: 38 dB, 58 dB, 78 dB, and 98 dB SPL. ITD tuning curves show peaks and troughs. Gray bar indicates the baseline level and s.d. of firing activity. Note that peaks of ITD tuning curves were increased and saturated around 78–98 dB SPL, while the trough firing rate was once increased but eventually decreased and became lower than the base line firing level at high SPL. (**b**) ITD tuning curves measured in 2400 Hz BF NL neurons. Firing rates were increased with the increase of SPL both at the peak and the trough of ITD tuning curve. (**c**) SON was electrolytically lesioned in the right side. Locations of NM, NL, and SON are indicated in the left side. From Nishino et al. (2008)

compensation of sound level effect on ITD tuning was confirmed by simulation in *NEURON* model (Nishino et al. 2008). In the simulation, it was assumed that (1) balanced excitatory synaptic inputs were made in NL from both sides of NM, (2) the power of SON inhibition was proportional to the SPL (sound pressure level), and (3) the inhibition operated as tonic (or sustained) inhibitory conductance during the

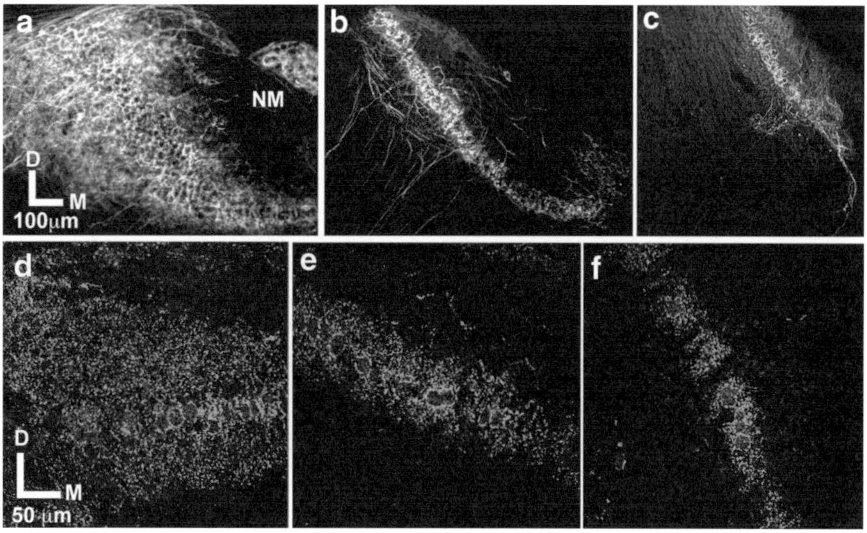

Fig. 2.26 SON projections to NL
(**a**, **b**, **c**) SON projections were labeled by fluorescence protein transfected in SON neurons by Sindbis virus. (**a**) low-CF; (**b**) middle-CF; and (**c**) high-CF regions. (**d**, **e**, **f**) VGAT, vesicular GABA transporter, was immuno-labeled in NL in low-CF (**d**), middle-CF (**e**), and high-CF (**f**) regions. Note that CF region-dependent pattern of the label was similar between SON projection and VGAT.
From Nishino et al. (2008)

input of sound signal. The simulation demonstrated a firing activity to occur within a narrow phase angle, and a good contrast of firing rate difference between the best and worst ITDs was confirmed as it was so in Fig. 2.27 (control).

The imbalance of bilateral excitatory inputs could be included in the neuron model to emulate the SON lesion, by setting the ipsilateral conductance twice the size of the contralateral conductance, while the average conductance of bilateral EPSPs was maintained constant (Nishino et al. 2008). The firing activity was well phase-locked at each ITD, but the phase was shifted with ITD throughout 360 degrees (similar to Fig. 2.27a lesion), and the firing rate was increased even at the worst ITD when a loud sound was applied. Thus, the peak-trough contrast of ITD tuning was critically dependent on the balance of bilateral excitatory inputs. SON receives sound level information from the ipsilateral nucleus angularis (NA) and issues the inhibitory signals to ipsilateral NM, NA, and NL and to the contralateral SON. These inhibitory projections operate as loops of negative feedback and balance the firing activity of two sides and apparently control the NL activity to enhance the tolerance of ITD processing to the sound level (Nishino et al. 2008).

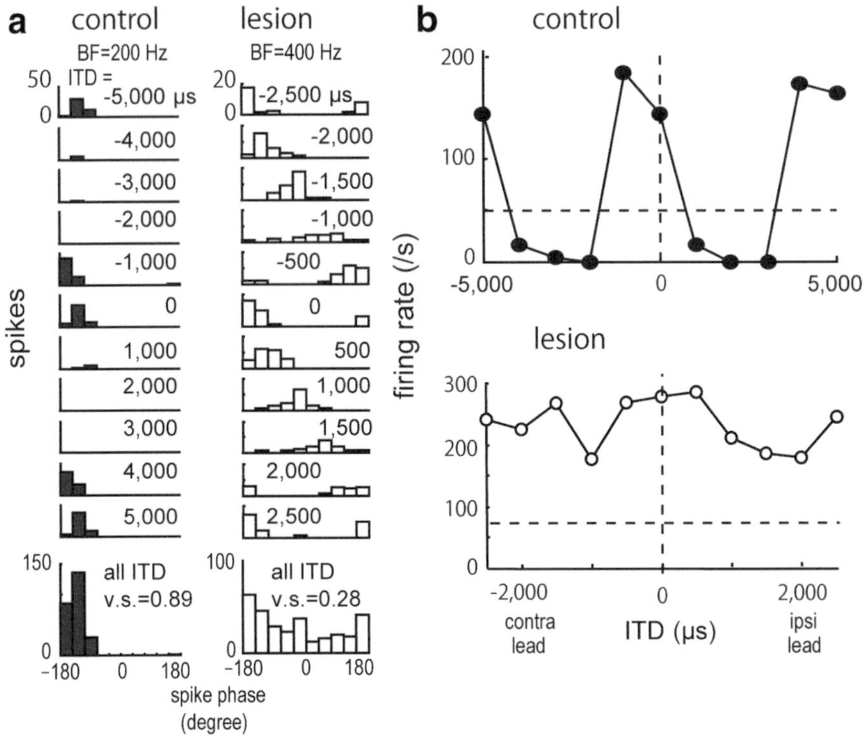

Fig. 2.27 SON activity modifies the phase-locked firing property of NL neurons
(**a**) phase-locked firings of NL neuron at different ITDs in the control with intact SON (left, BF 200 Hz) and after lesion of SON (right, BF 400 Hz). Phase of unit activity was fixed regardless of ITDs when SON was intact. SON lesioning affected the phase of firing, and the best phase was displaced with ITD. (**b**) ITD tuning in the control and after the SON lesion. ITD tuning curves were plotted from **a**.
From Nishino et al. (2008)

2.10 Sound Level-Dependent Inhibition Has Critical Roles to Tune the ITD Processing in Both Mammals and Birds

ITD tuning in MSO was also controlled by inhibition. MSO is a homologous nucleus in mammals to NL in birds and processes ITD from excitatory inputs from bilateral cochlear nuclei (Fig. 2.28). In the dog (Goldberg and Brown 1969) and the cat (Yin and Chan 1990), the firing rate of MSO units at the worst ITD was lower than the rate at monaural stimulation and sometimes even lower than the spontaneous firing rate. MSO receives inhibitory projections from ipsilateral lateral nucleus of the trapezoid body (LNTB) and from medial nucleus of the trapezoid body (MNTB). MNTB receives excitatory inputs from the globular bushy cells in the contralateral anteroventral cochlear nucleus (AVCN) (Cant and Hyson 1992), and

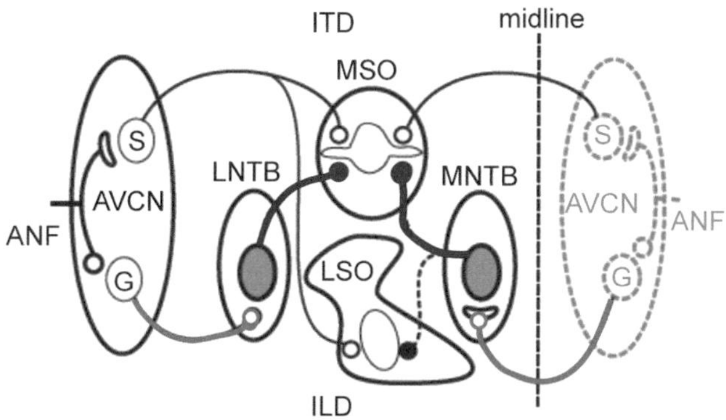

Fig. 2.28 Scheme of mammalian brainstem acoustic nuclei
MSO receives bilateral excitatory inputs (open circles) from AVCN and processes ITD. Both LNTB and MNTB receive excitatory inputs from globular bushy cells (G) of ipsilateral (LNTB) or contralateral AVCN (MNTB) and send inhibitory signals (filled circles) to MSO to modify the ITD processing. In LSO ILD is processed between the excitatory inputs from spherical bushy cells (S) in AVCN and the inhibitory inputs from MNTB

the firing activity is tightly phase-locked to the contralateral sound (Smith et al. 1998; Tollin and Yin 2005), implying a presence of precisely timed inhibitory inputs in MSO. These inhibitory inputs are feed forward inhibition and proposed to shape the time window of coincidence detection (Grothe and Sanes 1994; von Gersdorff and Borst 2002). Activity of inhibitory inputs to MSO from ipsilateral LNTB was dependent on the sound level (Smith et al. 1991). MNTB neurons provide inhibitory inputs to the lateral superior olive (LSO) where interaural level difference (ILD) is processed. Accordingly, sound level information of inhibitory nature is utilized in processing of the sound timing information in MSO. However, mammalian and avian species appear to take different strategies to use inhibitory information in processing of ITD; the feedforward inhibition seems adopted in mammals, and the feedback inhibition is adopted in birds (Brand et al. 2002; Grothe 2003; Harper and McAlpine 2004; Burger et al. 2005; Nishino et al. 2008). Feedforward inhibition may have advantages in speed and simpler network structure over the feedback inhibition, although the feedback inhibition in birds may operate more stably and seems appropriate for controlling the fine-tuning of ITD processing for the low-frequency sound.

2.10.1 Roles of Phase-Locked Inhibition in ITD Tuning of Mammals

The ITD tuning is greatly refined by inhibitory synapses in mammals (Goldberg and Brown 1969: Spitzer and Semple 1995; Yin and Chan 1990). In addition to excitatory inputs from bilateral AVCNs that are utilized to encode ITD, mammalian

MSO neurons receive inhibitory signals from bilateral sources: ipsilateral inhibitory signal through LNTB and the contralateral inhibitory signal through MNTB (Fig. 2.28). In vivo study conducted in the gerbil MSO reported that the inhibition from MNTB reached MSO preceding the contralateral EPSP and delayed the rising phase of EPSPs, despite a presence of one extra synapse in the transmission pathway (Fig. 2.28). The inhibition through the LNTB reached MSO after EPSP and accelerated the decay of ipsilateral EPSP (Brand et al. 2002). It was proposed that the delayed rising of contralateral EPSP displaced the maximum firing of MSO neuron to the contralateral leading ITD. Accordingly, the steep slope of the ITD tuning curve was placed within the physiologically relevant range of ITDs, and the ITD could be coded uniquely as the firing rate of the MSO neuron (slope coding) (McAlpine et al. 2001; Brand et al. 2002; Grothe 2003; Pecka et al. 2008). The acceleration of ipsilateral EPSP decay enhanced the sharpness of ITD tuning. However, there remains debates about the roles of leading IPSP from MNTB and on the shift of ITD tuning curve (Joris and Yin 2007).

Two important ideas were proposed in the hypothesis on the role of inhibition in ITD tuning of MSO (Brand et al. 2002; Grothe 2003; Pecka et al. 2008): (1) IPSP preceded EPSPs in the contralateral excitatory inputs only, and (2) the leading IPSP delayed the rise of EPSP, and the delay resulted in the displacement of ITD tuning curve to the contralateral leading ITD. However, several recent experiments provided results that are not consistent with this proposal (Joris and Yin 2007; Roberts et al. 2013; van der Heijden et al. 2013; Franken et al. 2015). By using thick slice preparations (1.0–1.5 mm thickness) of the gerbil brainstem that contains cochlear nucleus (CN), superior olivary complex (SOC), and the ANF entry to the brainstem, the electrical stimulation of ANF at the entry to the brainstem demonstrated that IPSPs arrived MSO earlier than EPSPs for both ipsilateral and contralateral stimulations (Robert et al. 2013). It was also demonstrated that IPSP shunted the membrane conductance and hyperpolarized the membrane. However, those effects of IPSPs on the coincidence detection were balanced by the deactivation of KLVA (Kv1) channel that occurred as a result of the hyperpolarization. The inhibition narrowed the time window for coincidence detection, but did not accompany the shift of ITD tuning curve. These results of in vivo whole-cell recordings in gerbil MSO demonstrated that the timing of inhibitory inputs from the contralateral ear was not sufficiently precise. The excitatory inputs from both ears were summated linearly, and the intrinsic membrane conductance boosted the spike firings when the summated depolarization was large enough (van der Heijden et al. 2013). Moreover, it was demonstrated by these in vivo whole-cell patch clamp recordings that the suppression of glycinergic IPSPs did not produce the expected shift of ITD tuning curve toward 0 μs ITD (Franken et al. 2015). The previous report on the shift of ITD tuning curve (Brand et al. 2002) was argued as a consequence of non-specific effects of strychnine used in high concentration in that experiment; strychnine is the inhibitor of glycinergic synapse. They further proposed an idea that the coincidence detection at MSO was not made by the instantaneous comparison of bilateral inputs, but it was made by the interaction between intrinsic membrane conductance such as Kv1 and

Na^+ channels and the synaptic activity. These observations are not consistent with the hypothesis that IPSP inputs displace the ITD tuning curve (Brand et al. 2002).

Because of the anatomical location of MSO in the ventral border of brainstem, access of recording electrode to MSO neurons was difficult. Furthermore, the presence of large acoustic field response made electrophysiological recording of spikes difficult in MSO. Even in successful cases of whole-cell or juxtacellular recordings, the fast time course of synaptic potentials and the small size of spikes deteriorated the quality of recordings in the MSO. The smallness of action potentials indicates generation of spikes at remote site, perhaps at the first node of Ranvier, and the back propagation of them to the soma where whole-cell recording was made. It seems required to invent some novel recording technology of improved signal to noise ratio and to challenge the high quality recordings both in vitro and in vivo experiments, for the understanding of fundamental mechanism of ITD tuning in the mammalian MSO.

2.10.2 Phase-Locked Inhibition of NL Through NM Interneuron

Low-CF NL neurons receive IPSCs from local GABAergic neurons in addition to the tonic (or sustained) inhibition from the SON (Yamada et al. 2013). This IPSC tracked a spike of ipsilateral NM neuron monosynaptically and limited the time window short for the summation of excitatory inputs. Accordingly, this is a feedforward inhibitory circuit and plays additional roles in processing ITDs of the low-frequency sounds in NL, as similar to the operation of inhibitory circuits in MSO of mammals (Peña et al. 1996; Dasika et al. 2005; Nishino et al. 2008; Pecka et al. 2008; Yamada et al. 2013). Consequently, the ITD tuning of NL neurons could be maintained over a wide range of sound pressure level.

In low-CF NL neurons in brainstem slices, PSCs (postsynaptic currents) of slower kinetics followed the fast PSCs, when the ipsilateral NM projection fibers were electrically stimulated (Fig. 2.29Aa, Yamada et al. 2013). Both currents were blocked by bath-applied DNQX (20 μM), antagonist of AMPA receptors. However, only the slower component was eliminated by bicuculline (20 μM) or gabazine (10 μM), antagonists of $GABA_A$ receptors (Fig. 2.29Ab). Therefore, the fast component was an EPSC, and the slower component was a GABAergic IPSC. Here, GABAergic IPSC was inward going because the Cl^- composition of recording medium was high due to the use of KCl-based internal medium in the experiment. In high-CF and middle-CF neurons, these IPSCs were not found. Furthermore, these IPSCs were not generated by stimulation to the contralateral NM projection. Every IPSC followed the EPSC with a delay of 1–2 ms when the stimulus intensity was sufficiently high and the temporal jitter of IPSCs was small (0.21 ± 0.04 ms). The time-locked properties of IPSC to EPSC were not affected by tetanus frequency from 100 to 333 Hz (Fig. 2.29B). These results indicated that local GABAergic

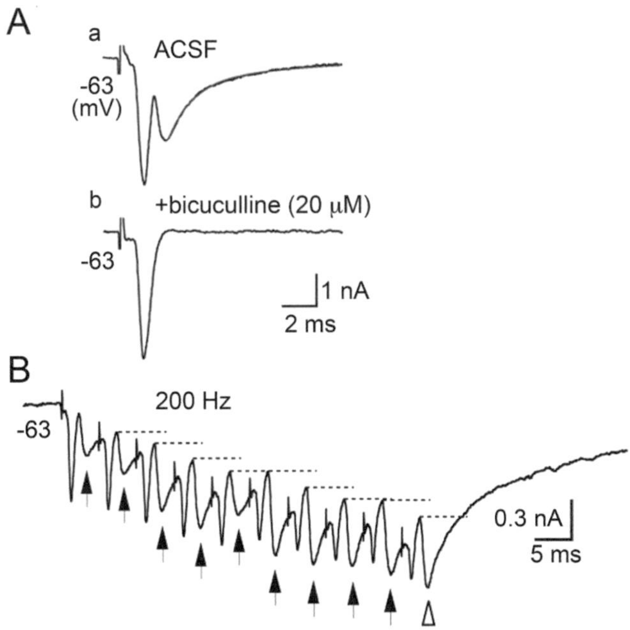

Fig. 2.29 IPSC that follows ipsilateral NM activity in low-frequency NL neurons
(A) electrical stimulation applied to the NM projection fiber evoked biphasic synaptic currents in ipsilateral low-frequency NL neurons (Aa) in slice preparations that contained both NM and NL. Second synaptic current was GABAergic, since bicuculline (20 μM) eliminated the late current (Ab). DNQX (20 μM) suppressed both currents. Recordings were made with a patch electrode filled with KCl-based internal medium; thus, both EPSC and IPSC were inward going in A. (B) biphasic synaptic currents followed high-frequency electrical stimulation
From Yamada et al. (2013)

neurons tracked the ipsilateral NM spikes. It was likely that NM neurons innervate the local GABAergic neurons monosynaptically.

Using in situ hybridization, many GAD1-positive cells were found in the low-CF region of NL and NM. GAD, glutamic acid decarboxylase, is a synthesizing enzyme of GABA. However, GAD1-positive cells were rarely observed in the high-CF and middle-CF NL regions. The GAD1-positive neurons were clustered in the area between NM and NL, within NL and lateral to NL. Dense GABA immunoreactivity was found in fibers and terminals in NL, while GABA immuno-positive cell bodies were clustered in the lateral border of NL. These distributions were consistent with previous reports describing clusters of GABAergic neurons in these areas (Carr et al. 1989; von Bartheld et al. 1989). Nearly 80% of the GAD1-positive cells were found in the low-CF NL region.

Because these GABA neurons were apparently activated by glutamatergic synapse from NM (see Fig. 2.29), Yamada et al. (2013) focally activated GABAergic neurons by using a caged-glutamate compound outside the low-CF NL region. UV illumination uncaged glutamate and evoked transient synaptic currents in low-CF

NL neurons. These UV-activated currents were blocked by bath-applied bicuculline. Thus, local GABAergic neurons likely formed inhibitory synapses on low-CF NL neurons.

IPSCs were activated at precise timing after the spike of NM neurons (Fig. 2.29B). The GABA receptor subtype that can be activated with fast kinetics should be involved in the PSC. Spontaneous IPSCs measured in low-CF NL neurons exhibited faster decay than those measured in the middle-CF and high-CF NL neurons. The average decay time constant was approximately four times smaller in the low-CF NL neurons than in the high-CF and middle-CF NL neurons. Subunit composition is known to affect the kinetics of $GABA_A$ receptors (Mody and Pearce 2004; Farrant and Nusser 2005). Presence of α1-subunit in $GABA_A$ receptors accelerated the decay kinetics of IPSC (McClellan and Twyman 1999; Vicini et al. 2001; Goldstein et al. 2002; Eyre et al. 2012). In situ hybridization demonstrated a relatively high-level expression of mRNA encoding $GABA_A$ α1 receptor subunit in the caudo-lateral NL region (the low-CF NL region). The expression diminished toward the rostro-medial NL region where neurons receive high- to middle-frequency inputs. Accordingly, the fast IPSCs in low-CF NL neurons are attributable to the activation of α1-subunit of $GABA_A$ receptors. In contrast, expression of α1 $GABA_A$ receptor subunit was not observed in NM.

2.10.3 Cooperation of Tonic Inhibition and Phasic Inhibition to Enhance the Tolerance to Sound Level

Based on the hypothesis that local GABA neurons produce a phase-locked inhibition on low-CF NL neurons, we simulated on a *NEURON* model how this phasic inhibition affects ITD tuning (Yamada et al. 2013). In low-CF NL neurons, IPSC was set to delay a fixed short time to ipsilateral EPSC and to decay quickly. This phasic inhibition likely resets the summation of synaptic inputs shortly after the input of ipsilateral EPSC and narrows the time window for the summation of bilateral EPSPs.

Two modes of inhibitory inputs were introduced to the soma in the model NL neuron: a tonic inhibition that mimicked the inhibition from SON neurons and a phasic inhibition from local GABAergic neurons that tracked ipsilateral NM spike. The phasic IPSC was assumed to delay 2 ms to the ipsilateral EPSC. Parameters were set to induce spikes when bilateral sound inputs were made in the absence of inhibitory inputs. Without any inhibition action potentials were generated even by unilateral EPSPs, particularly when high-level excitatory inputs (90 nS, Fig. 2.30a, d) were made. Thus, the firing rate was apparently higher at the out-of-phase ITD stimulus than at the in-phase ITD stimulus (red open square, Fig. 2.30g).

When inhibition of either mode was included in that model neuron, a sharp contrast of firing activity was found between the in-phase and out-of-phase inputs in two cases. One was when weak excitatory inputs were made under the phasic inhibition (Fig. 2.30c, i-blue line). The other was the case of strong excitatory inputs under the tonic inhibition (Fig. 2.30e, h-red line). In the simulation, we separately introduced two modes of inhibition. During low-level or weak excitatory inputs (60 nS, Fig. 2.30b, c), the firing rate was low at the out-of-phase inputs both for the tonic inhibition (Fig. 2.30b) and for the phasic inhibition (Fig. 2.30c). Firing rate at the in-phase ITD was still low during the tonic inhibition (Fig. 2.30b). However, the firing activity was robust at the in-phase ITD under the phasic mode of inhibition (Fig. 2.30c). When the input signal was strong (90 nS), the phasic mode of inhibition was not effective at all to reduce the firing activity (Fig. 2.30f). Accordingly, under the phasic inhibition, the contrast of peak to trough firing rates was high only when the low-level or weak excitatory inputs were made (Fig. 2.30i-blue line). The tonic mode of inhibition was effective to reduce the firing rate at the out-of-phase inputs even during the high-level inputs (90 nS), while the firing activity was robust at the in-phase inputs (Fig. 2.30e). Accordingly, under the tonic inhibition, the contrast of ITD tuning was large when strong excitatory inputs were made (Fig. 2.30h-red line).

When the input excitatory conductance was systematically changed and the peak-trough contrast of ITD tuning function was monitored at several levels of inhibitory conductance, the contrast was enhanced by the tonic inhibition when the input excitation conductance was large (Fig. 2.31a) and by the phasic inhibition when the input conductance was small (Fig. 2.31b) (Yamada et al. 2013). When both tonic inhibition (×3 of the excitatory conductance) and phasic inhibition (180 nS, moderate level of inhibition) were included in the simulation, a large difference (threefold) of excitatory inputs did not affect much the ITD tuning function (Fig. 2.31c). By the combination of phasic and tonic inhibitions, the peak-trough contrast was increased at lower level of excitatory conductance inputs and was maintained high over wider ranges of excitatory conductance inputs than the case of tonic inhibition alone (Fig. 2.31d, black solid line is for the tonic inhibition alone; red lines are for the combinations of tonic and phasic inhibitions; the conductance of phasic inhibition was modified as indicated by different symbols). Overall, the cooperative activity of phasic inhibition and tonic inhibition made the contrast of ITD tuning curves tolerant to a wide range of excitatory conductance or the input sound level.

2.11 Sound Level Coding Pathway: Nucleus Angularis and Lemniscal Nuclei

Temporal information and level information of sound are transmitted to the brainstem auditory nuclei in parallel pathways (Fig. 2.1). However, the interplay between them is important in processing the interaural information (Olsen et al. 1989). In order to understand the property of neurons that transmit sound level information, we recorded from neurons in the nucleus angularis (NA) and in the lateral lemniscus (LL).

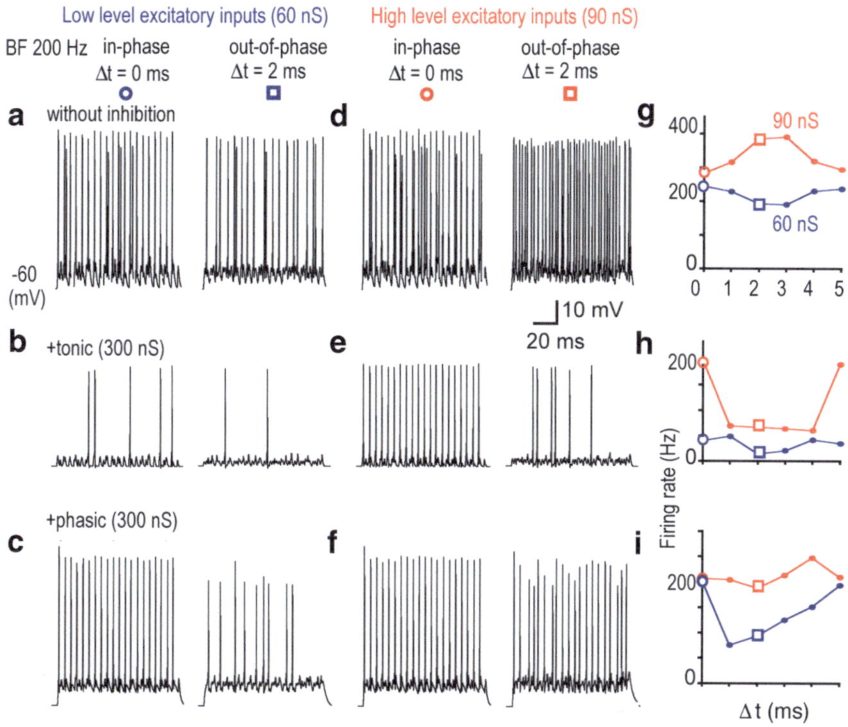

Fig. 2.30 Simulation of ITD tuning in a *NEURON* model of NL
Two types of inhibition were applied separately in a model neuron of NL under a low-level (60 nS) excitatory conductance inputs or a high-level (90 nS) excitatory conductance inputs. (**a**, **b**, and **c**) are the simulation of spike activity for the low-level excitatory inputs and (**d**, **e**, and **f**) for the high-level excitatory inputs. a and **d** are for neurons without inhibitory inputs, **b** and **e** are the cases for tonic (or sustained) inhibition, and **c** and **f** are the cases for the phasic inhibition. Bilateral excitatory inputs were made with zero-time difference (Δt, in-phase) or with 2 ms time difference (Δt, out-of-phase). Representative trace is shown in each condition. The resting potential was set at −60 mV. When inhibitory synapses were not included (**a**, **d**), robust firing activity occurred both in the low-level and the high-level excitatory conductance inputs. The firing rate was even higher at the out-of-phase inputs of high-level excitatory conductance inputs (**g**). When tonic inhibitory conductance was included (**b**, **e**, 300 nS), firing rate was depressed at the low-level excitatory inputs for both in-phase and out-of-phase stimulus (**b**) and for the out-of-phase stimulus of high-level excitatory conductance (**e**). The firing activity was robust at the in-phase inputs of the high-level excitatory conductance. Thus, the contrast of firing rates between in-phase and out-of-phase inputs was the largest at the high-level excitatory conductance (**h**). When a phasic inhibition was included (**c**, **f**, 300 nS), firing rate was suppressed only during the out-of-phase inputs of low-level excitation (**c**), and the contrast of firing rates was the largest for the low-level excitatory inputs (**i**). (**g**, **h**, and **i**) are plots of the firing rates as a function of Δt for both low-level (60 nS excitatory conductance, blue lines) and high -level (90 nS excitatory conductance, red lines) inputs
From Yamada et al. (2013)

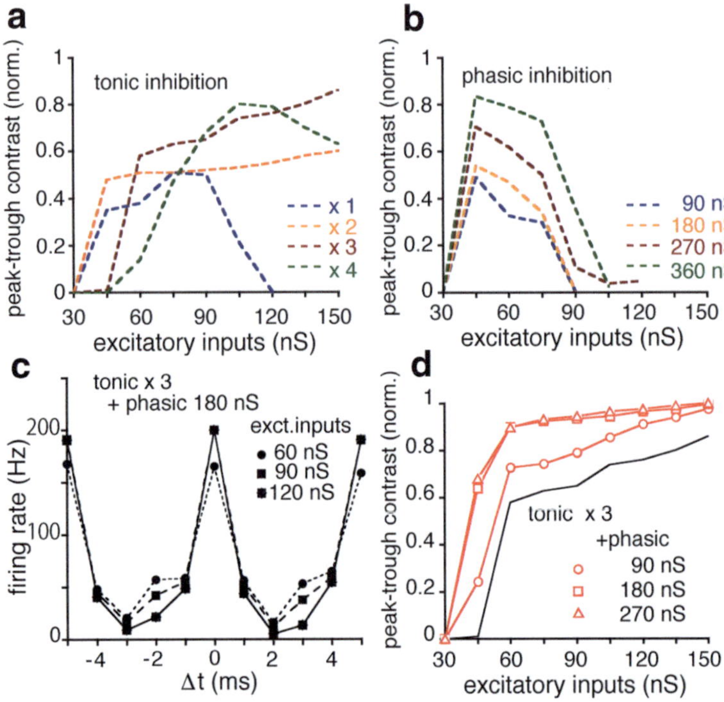

Fig. 2.31 Simulation of ITD contrast in neuron model
(**a**) peak-trough contrast (ratio between a peak firing rate and a trough firing rate) from an NL neuron model that receives only the tonic (or sustained) inhibition. Conductance of tonic inhibition was assumed from one to four times of the conductance of excitatory inputs. The conductance of excitatory inputs was changed, and peak-trough contrast of ITD tuning curve was plotted after normalization. Peak-trough contrast increased when the inhibition level was high. (**b**) peak-trough contrast was calculated from a phasic inhibition model. The level of phasic inhibition was assumed from 90 nS to 300 nS, and the conductance of excitatory inputs was changed as in **a**. The peak-trough contrast was maximum when the excitatory conductance input was small. (**c**) ITD tuning curves when both tonic inhibition and phasic inhibition were included. Conductance of excitatory inputs was changed from 60 to 120 nS; tonic inhibition was included as ×3 of the excitatory conductance inputs, while the phasic inhibition was constant at 180 nS. (**d**) peak-trough contrast was plotted as a function of excitatory conductance for four cases. When the phasic inhibition was included, the peak-trough contrast was increased at the small excitatory conductance inputs more than the level when the tonic inhibition alone was included (×3 of excitatory conductance, black line)
From Yamada et al. (2013)

2.11.1 Nucleus Angularis (NA)

In the chicken embryo (E16–E21), NA neurons fired in three patterns when current injections were made in slice preparations (Fukui and Ohmori 2003): onset pattern (19%), tonic pattern (52%), and the pattern of tonic firing with a pause after onset activity (29%). The pattern of tonic firing with a pause was likely attributed to the immaturity of membrane excitability, since neurons fired either onset (17%) or tonic patterns (83%) after hatch (P0–P5). After hatch, firing activity became robust: the threshold potential became lowered, the maximum firing frequency was increased, and both maximum rate of rise and fall of spikes were accelerated three to four times. The firing frequency of tonic neurons reached a maximum rate of about 650 Hz. Application of TEA (1 mM, tetraethyl ammonium) reduced the firing frequency, broadened individual action potentials, and reduced the maximum rate of fall, but the threshold current was not changed. TEA (1 mM) blocks the high-voltage-activated potassium channel Kv3 (KHVA). Currents through KHVA are required to fire at high rates (Grissmer et al. 1994; Rathouz and Trussell 1998; Wang et al. 1998). In contrast, dendrotoxin (DTX, 100 nM) blocked KLVA, reduced the threshold current, and made onset neurons to fire tonic.

Branching patterns of auditory nerve fibers (ANFs) in NA were visualized by labeling with DiI (1,1′-dioctadecyl-3,3,3′,3′-tetramethylindocarbocyanine perchlorate) that was placed within the cochlea organ (Fig. 2.32). DiI placed near the apex of cochlea (low-frequency region) labeled the ventral part of NA and DiI placed in the base (high-frequency region) labeled the dorsolateral part. Therefore, the axis of tonotopy in NA was directed from ventral low-frequency to dorsolateral high-frequency regions (Fig. 2.32). Neurons were further labeled with biocytin after measurement of firing activity in slices. Most labeled neurons were tonic type and extended dendrites in parallel with the ANF projection (Fig. 2.33). Thus, ANF activity of a limited range of frequency was likely extracted as the tonic activity of these NA neurons. Two onset neurons were encircled in Fig. 2.33 (Fukui and Ohmori 2003).

2.11.2 IPD Dependence of Sound Level Processing in NA

When NA unit was recorded in vivo, the firing activity increased monotonically with the increase of both ipsilateral sound and contralateral sound level (open circles, ipsilateral sound; filled circles, contralateral sound; Fig. 2.34a), although a higher-level contralateral sound was required to activate the NA neuron. This is because of an acoustic communication through the interaural canal that is formed

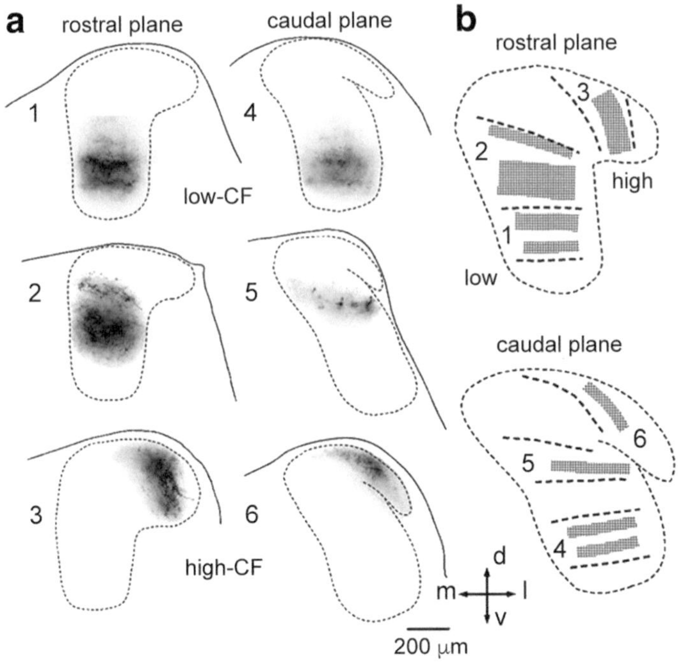

Fig. 2.32 Pattern of ANF projection to NA
(**a**) DiI crystal was placed in a cochlea organ. NA regions were sliced, and ANF projection was visualized as fluorescence images. Region of ANFs projection in NA was different dependent on the depth in cochlea organ where DiI crystal was placed. When DiI was placed in the apex (low-CF region), the ventral part of NA was labeled, and the dorsal part of NA was labeled when DiI was placed in the base of cochlea (high-CF region). (**b**) schematic illustrations to show the labeled region in the NA. Numbers are corresponded to the number of slices in **a**
From Fukui and Ohmori (2003)

between two middle ear cavities (Fig. 2.34b). Extra impedance was included in the transmission of sound through the interaural canal, so that SPL (sound pressure level) of the contralateral sound was higher about 16 dB than the ipsilateral sound in order to evoke the same level of activity of NA neurons; however, the slope of rate intensity relationship and the maximum level of firing rate were not different between the stimulus of two sides (Fig. 2.34a).

When binaural sound was applied in-phase and the contralateral sound level was changed while keeping the ipsilateral sound level constant (52 dB SPL), the firing rate of NA neuron was reduced significantly at certain level of the contralateral sound (filled diamonds, Fig. 2.35). The firing rate decreased sharply between 40 and 65 dB SPL; then the firing rate was increased with the increase of contralateral sound level. There was a dip between the falling and the rising parts of the rate intensity function (arrow, Fig. 2.35). The sound pressure level difference at the dip was 14 dB SPL in this unit and was close to the level difference of rate-level functions in Fig. 2.34a, when sound stimulus was applied separately to two ears. When

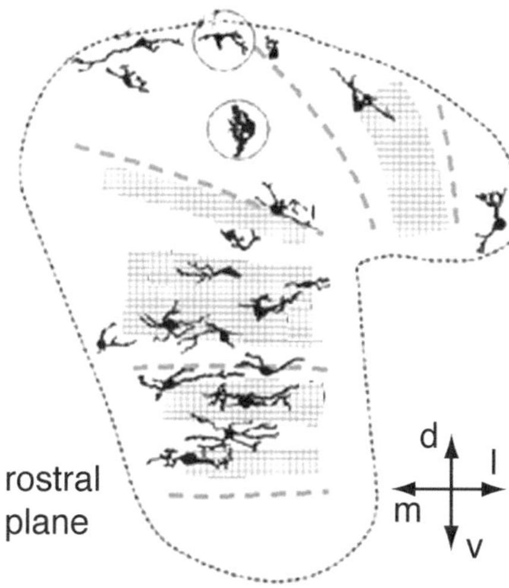

Fig. 2.33 Branching pattern of the dendrites of NA neurons
NA neurons were biocytin labeled in slices. Shaded boxes indicate the pattern of projection of ANFs in the rostral plane of NA (Fig. 2.32b). Note that most neurons extended dendrites parallel to the ANF projections in the nucleus. Most labeled neurons are tonic firing type, and two onset neurons are encircled
From Fukui and Ohmori (2003)

contralateral sound was applied alone, the rate-level function increased monotonically (open circles, Fig. 2.35).

More interestingly in an NA neuron of 200 Hz BF, the phase of spike firing was displaced at certain difference of sound level when the sound stimulus was applied in-phase to two ears (Fig. 2.36). Here the contralateral sound was increased gradually from 10 dB to 90 dB SPL while maintaining the ipsilateral sound at 63 dB SPL. The phase angle of unit activity was 170 degrees when the contralateral sound level was low (10–70 dB SPL) and was displaced to 310 degrees when the contralateral sound level was high (90 dB SPL). The neural activity was minimum at the contralateral sound level of 80 dB SPL (binaural difference of 17 dB SPL). This phase displacement was introduced by the interference of binaural sound waves within the interaural canal (IAC).

Certainly, the interference within the IAC was affected by the phase difference of bilateral sounds (Fig. 2.37). The depth of dip in the rate-ILD function was clearly affected by IPD (interaural phase difference). The dip was the deepest when the bilateral sound was in-phase (0 degree IPD), and the dip disappeared when the bilateral sound was out-of-phase (180 degrees IPD) in a neuron of BF 200 Hz (Fig. 2.37a). In neurons of higher BF, the phase shift was not clear in the unit activity, but the depth of dip was still affected by IPD in the rate-level function (Sato et al. 2010). In a neuron of BF 2540 Hz (Fig. 2.37b), the deepest dip occurred at the contralateral leading phase approximately by 60 degrees, but not at 0 degree IPD. This IPD was different from the dip IPD of low BF neurons (0 degree IPD). This shift of dip IPD

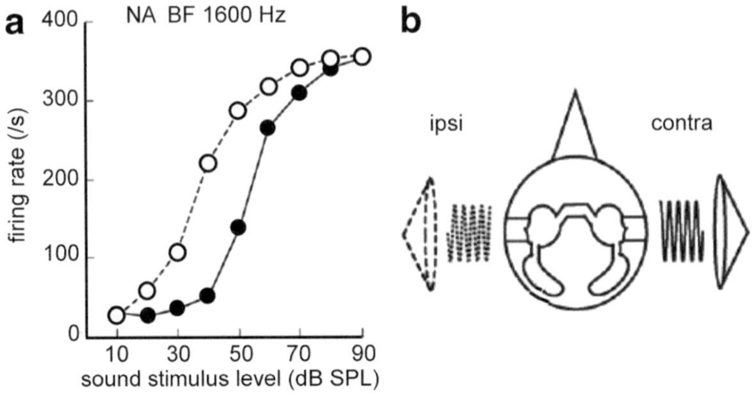

Fig. 2.34 Rate-level function of an NA neuron in vivo
(**a**) both ipsilateral (open circles) and contralateral (filled circles) sounds evoked firing responses in an NA neuron of BF 1600 Hz. (**b**) scheme of interaural canal
From Sato et al. (2010)

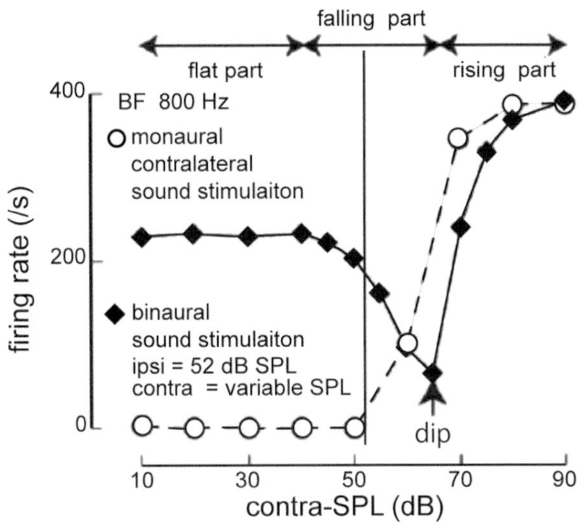

Fig. 2.35 Firing activity of an NA neuron to binaural sound stimuli
Bilateral sound of BF 800 Hz was applied in-phase (◆) by changing the SPL of contralateral sound, while the ipsilateral sound level was maintained constant at 52 dB SPL. The rate-level function was changed with the increase of contralateral sound level. The whole function was consisted of a flat part, a falling part and a rising part. Upward arrow indicates a dip where the firing rate was minimum. The dip occurred at 66 dB SPL of contralateral sound or at the interaural level difference (ILD) of 14 dB. The monaural response to the contralateral sound is indicated by open circles (O)
From Sato et al. (2010)

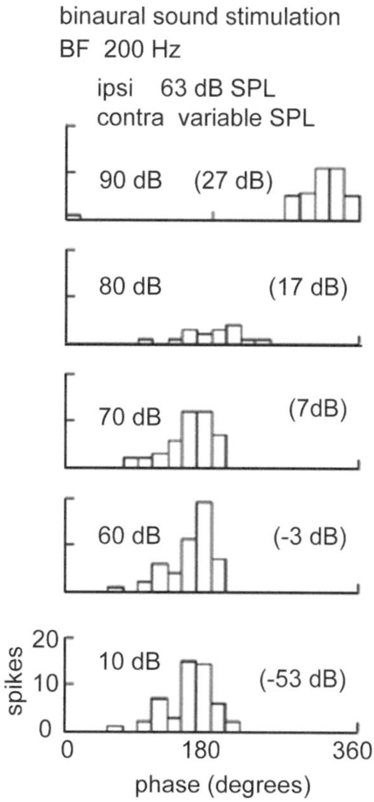

Fig. 2.36 Period histogram of firing activity of an NA neuron to binaural sound
When in-phase binaural sound of 200 Hz BF was applied, the phase of firing activity was changed depending on the ILD. The ipsilateral sound level was constant at 63 dB SPL, while the contralateral sound level was changed. The phase of neural activity was about 170 degrees when the contralateral SPL was low and was shifted by about 180 degrees when the contralateral SPL was high. ILD was indicated in brackets. The firing activity was shifted from the phase-locked to the ipsilateral sound to the one phase-locked to the contralateral sound as the increase of contralateral SPL. The firing activity was minimum at ILD of the phase transition (17 dB) that corresponded to the dip ILD in Fig. 2.35
From Sato et al. (2010)

should be attributed to the travel time of sound through the IAC. The travel time was estimated as 50–75 μs (Hyson et al. 1994). For low-frequency sound, the extra travel time was negligible because of a long cycle time (5000 μs at 200 Hz, Fig. 2.37a), and also the extra phase introduced (3.6–5.4 degrees) was negligible. However at higher frequencies (cycle time 394 μs at 2540 Hz, Fig. 2.37b), the travel time introduced extra phase difference (45–68 degrees) that was not negligible, and the dip-IPD was displaced significant from 0 degree (Sato et al. 2010).

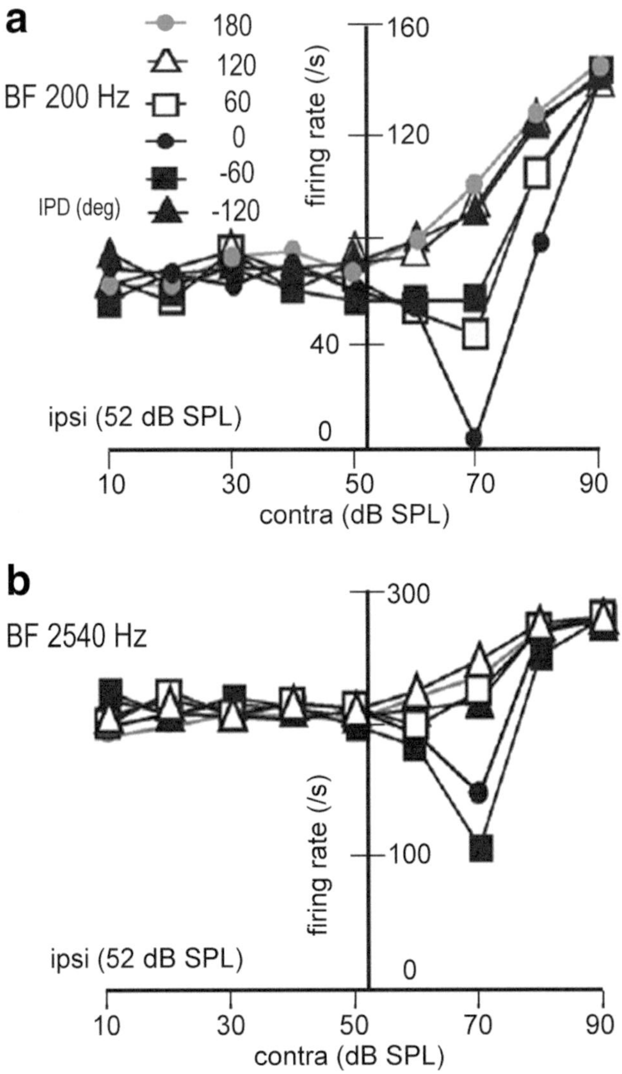

Fig. 2.37 Rate ILD functions of NA units were affected by interaural phase difference (IPD) of sound

(**a**) firing activity at the dip was minimum in an NA neuron of BF 200 Hz, when bilateral sound was applied in-phase. The depth of dip decreased with the increase of sound phase difference between two ears. (**b**) the largest dip occurred at slightly different sound phase (IPD of −60 degrees) from the in-phase (0 degree) sound in the NA neuron of BF 2540 Hz

From Sato et al. (2010)

2.11.3 Posterior Part of Lateral Lemniscus Dorsalis LLDp

Neurons in the dorsal lateral lemniscal nucleus (LLD), especially the neurons in the posterior part of LLD (LLDp), respond to interaural level difference (ILD) (Takahashi and Konishi 1988; Mogdans and Knudsen 1994). LLDp neurons receive excitatory inputs from NA (Fig. 2.1). In contrast to the bilateral response property of NA neurons (Fig. 2.34a), LLDp neurons responded only to the contralateral sound. Ipsilateral sound did not generate spikes (Fig. 2.38a). This is because of inhibitory connections between bilateral LLDp (Fig. 2.38b). Interestingly the frequency selectivity was the same in LLDp units between the excitatory input from contralateral NA and the inhibitory input from the contralateral LLDp. BFs of inputs were well matched between the excitatory and inhibitory inputs in these LLDp units (n = 20 units) (Manley et al. 1988; Sato et al. 2010). This mutual inhibition is the basis of ILD coding in the LLDp. Accordingly, the firing rate of LLDp units was increased with the increase of contralateral sound level while decreased with the increase of ipsilateral sound level (Fig. 2.39a). The rate-level function of LLDp units was determined by ILD (defined here as ipsilateral SPL-contralateral SPL) (Fig. 2.39b).

As expected from the IPD sensitivity of NA unit activity, the rate-ILD function of LLDp units was also IPD sensitive (Fig. 2.40). There are two patterns of IPD sensitivity in the rate-ILD functions. Different patterns are depended on the balance between the contralateral excitation and ipsilateral inhibition. When the inhibition was strong, the rate-ILD functions were displaced toward negative ILD (larger contralateral SPL), and the firing activity in positive ILDs (larger ipsilateral SPL) was

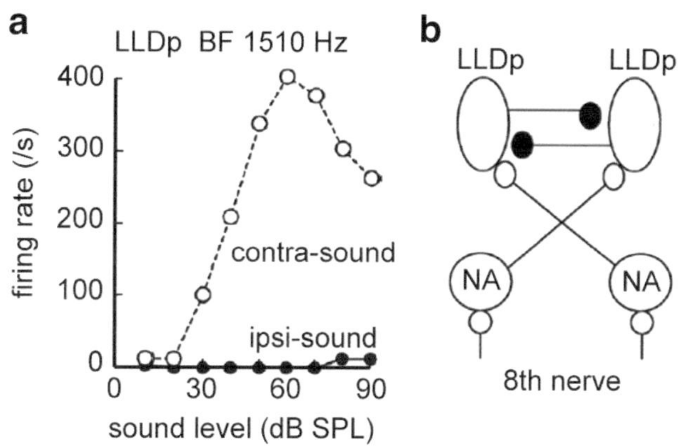

Fig. 2.38 Rate-level function of an LLDp unit to monaural sound
(**a**) rate-level functions of BF 1510 Hz LLDp unit when monaural sound stimulus was applied, either contralaterally (○) or ipsilaterally (●). (**b**) LLDp neurons receive excitatory inputs from contralateral NA neurons. Ipsilateral NA activity generated inhibitory synaptic inputs in LLDp units through the activation of inhibitory neurons in the contralateral LLDp
From Sato et al. (2010)

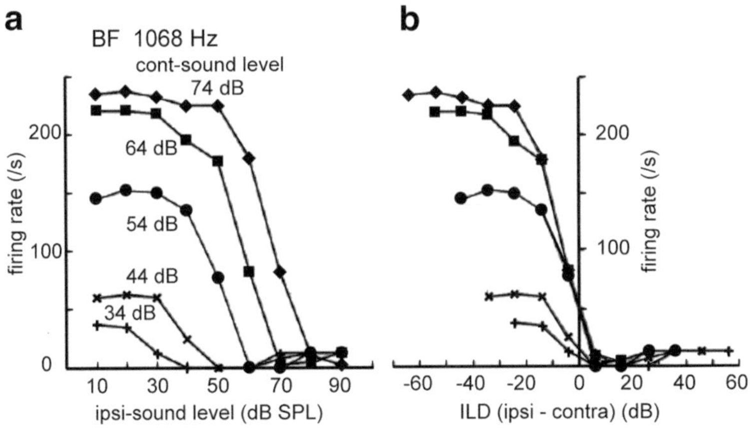

Fig. 2.39 Rate-level function of LLDp unit to binaural sound
(**a**) LLDp unit was responded to binaural sound of BF 1068 Hz. Contralateral sound level was fixed at five different levels from 34 to 74 dB as indicated, while the ipsilateral sound level was changed. Firing rate was plotted as a function to ipsilateral sound level. (**b**) firing rates in **a** are replotted against ILD
From Sato et al. (2010)

almost negligible in one group of LLDp units (Fig. 2.40a). Although the dip was not present in the rate-ILD function, the IPD effects were still present. The rate-ILD function was located most negative when the bilateral sound was applied in-phase (0 degree IPD). With certain IPDs the midpoint of rate-ILD function was displaced toward 0-dB ILD, and the firing rate was generally increased (Fig. 2.40a). In the other group of LLDp units, the inhibition was relatively weak; the rate-ILD functions demonstrated ILD sensitivity at more positive ILDs (Fig. 2.40b) than the former group (Fig. 2.40a). The firing rate changed almost linearly around 0-dB ILD for small ILD changes, and the dip was present at positive ILD in the rate-ILD functions (Fig. 2.40b). The depth of dip was dependent on IPD and was deepest when binaural sound was applied nearly in-phase.

This IPD dependence of rate-ILD function of LLDp neurons would enhance the ILD sensitivity to the sound source displacement, particularly in the contralateral hemifield. Sato et al. (2010) evaluated the ILD sensitivity when IPD was included in the binaural sound (Fig. 2.40) based on several observations: (1) any displacement of the sound source from the front of head (0-dB ILD) generates IPD in addition to ILD, and (2) the sound source movements toward the contralateral hemifield from the front of head would increase the firing rate as measured in the ordinary ILD-tuning function (Fig. 2.39b). The corresponding change of IPD from 0 degrees further increases the firing rate and the steepness of rate-ILD function (Fig. 2.40a) or at least the firing rate (Fig. 2.40b). (3) In contrast, the sound source movement toward the ipsilateral hemifield reduces the firing rate. The firing rate at 0-dB ILD is already at the minimum level and remains low in positive ILDs in one group of LLDp units (Fig. 2.40a). Thus, the firing rate will not be further affected by IPD in these LLDp units. In other group of LLDp units (Fig. 2.40b), the firing rate is high

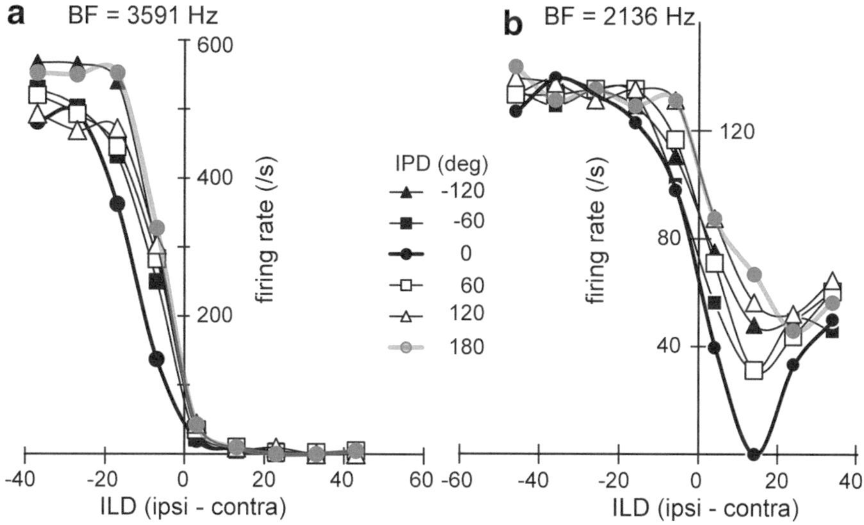

Fig. 2.40 Two types of IPD dependence of rate-ILD function of LLDp units
(**a**) 3591 Hz BF LLDp unit. The slope of rate-ILD functions is in the contralateral dominated ILD. There is no clear dip in these rate-ILD functions, but IPD dependence was remained as the displacement of the function along the ILD axis slightly toward right. The rate-ILD function was located most left when binaural sound was applied in-phase. (**b**) BF 2136 Hz LLDp unit. Slopes of rate-ILD functions are located around 0-ILD, and clear dip-ILDs are observed in all rate-ILD functions. The dip was the deepest when nearly in-phase sound was applied, and the depth of dip decreased with the increase of phase difference between the sound applied to two ears
From Sato et al. (2010)

at 0-dB ILD and will be changed across 0-dB ILD almost linearly as a function of ILD. Thus, the sound source displacement to ipsilateral hemifield would reduce the firing rate significantly. However, the accompanied change of IPD will increase the firing rate and will compensate for the reduced level of firing activity in these LLDp units (Fig. 2.40b). (4) Such IPD dependence of the rate-ILD functions of LLDp neurons would increase the ILD sensitivity in the contralateral hemifield more than that expected in the absence of IPD effect. Accordingly, the sound level information and temporal information interact each other for the processing of ILD in LLDp neurons (Fig. 2.40). We have described a similar interaction between the sound temporal information and the level information in ITD processing in the NL (Fig. 2.25).

2.12 Recapitulation of this Chapter

Membrane excitability, synaptic transmission, and network activity develop in parallel during maturation of animals and have critical importance in the brainstem auditory function. As for the membrane excitability, the low-voltage-activated K^+ channel has the key importance. Activation of this channel by small membrane depolarization accelerates EPSPs and action potentials. This is critical to achieve the temporal precision of auditory signal. Na^+ channels are arranged in most effective way to generate spikes in the axon initial segment. Synapses are arranged strategically to reduce the temporal fluctuation of transmission. High-frequency signals are transmitted by utilizing large EPSCs that are essential to achieve a high-fidelity transmission in NM, and low-frequency signals are transmitted by summation of many small EPSCs to reduce temporal jitter of spike generation. The progress of Na^+ channel inactivation during slow depolarization by the summation of small EPSPs could be overcome by accumulation of a large number of Na^+ channels in the axon initial segment in the low-frequency NM neurons. Displacing the axon initial segment to a little remote location from the cell body also overcomes the Na^+ channel inactivation during high-frequency synaptic inputs in the high-CF NL neurons. Also, the reduced expression or absence of Na^+ channel expression in the soma of NM and NL neurons practically prevented the cell soma from spiking and accelerated the spike conduction in the axon by reducing the capacitive load of cell soma.

ITD coding in NL is further modulated through the network activity of inhibitory neurons in SON or through the phase-locked inhibition from NM local neurons. Activity of these inhibitory neurons makes the ITD processing tolerable to a wide range of sound level. ITD tuning is refined by the involvement of inhibitory circuits in both mammals and in birds. Although, some differences are present in the operation pattern of inhibitory circuits between mammalian MSO and avian NL, such as the feedforward inhibition in the MSO and the feedback inhibition in the NL, in both species the signal processing of ITD is modulated by inhibitory circuits, and the dynamic range of coincidence detector neurons is increased by the inhibitory circuits in the sound level dependent manner.

The sound level signal is processed in a separate pathway from the sound temporal signal. However, the processing of temporal information and level information is mutually affected. The processing of ITD is affected by sound level signal through the inhibitory circuits, and the sound level-sensitivity is affected by interaural phase difference. The phase difference enhances the direction sensitivity of sound level-sensitive neurons. Accordingly, both temporal and level signals of sound interact each other to enhance precision of the sound source localization.

References

Araki T, Otani T (1955) Response of single motoneurons to direct stimulation in toad's spinal cord. J Neurophysiol 18(5):472–485

Banks MI, Pearce RA, Smith PH (1993) Hyperpolarization-activated cation current (Ih) in neurons of the medial nucleus of the trapezoid body: voltage-clamp analysis and enhancement by norepinephrine and cAMP suggest a modulatory mechanism in the auditory brain stem. J Neurophysiol 70(4):1420–1432

Barnes-Davies M, Forsythe ID (1995) Pre- and postsynaptic glutamate receptors at a giant excitatory synapse in rat auditory brainstem slices. J Physiol 488(Pt 2):387–406

Bender KJ, Trussell LO (2012) The physiology of the axon initial segment. Annu Rev Neurosci 35:249–265

Bobker DH, Williams JT (1989) Serotonin augments the cationic current Ih in central neurons. Neuron 2(6):1535–1540

Brand A, Behrend O, Marquardt T, McAlpine D, Grothe B (2002) Precise inhibition is essential for microsecond interaural time difference coding. Nature 417(6888):543–547

Brenowitz S, Trussell LO (2001) Maturation of synaptic transmission at end-bulb synapses of the cochlear nucleus. J Neurosci 21(23):9487–9498

Brückner S, Hyson RL (1998) Effect of GABA on the processing of interaural time differences in nucleus laminaris neurons in the chick. Eur J Neurosci 10(11):3438–3450

Burger RM, Cramer KS, Pfeiffer JD, Rubel EW (2005) Avian superior olivary nucleus provides divergent inhibitory input to parallel auditory pathways. J Comp Neurol 481(1):6–18

Calford MB, Piddington RW (1988) Avian interaural canal enhances interaural delay. J Comp Physiol 162:503–510

Cant NB, Hyson RL (1992) Projections from the lateral nucleus of the trapezoid body to the medial superior olivary nucleus in the gerbil. Hear Res 58(1):26–34

Carr CE, Fujita I, Konishi M (1989) Distribution of GABAergic neurons and terminals in the auditory system of the barn owl. J Comp Neurol 286(2):190–207

Carr CE, Konishi M (1990) A circuit for detection of interaural time differences in the brain stem of the barn owl. J Neurosci 10(10):3227–3246

Carr CE, Boudreau RE (1991) Central projections of auditory nerve fibers in the barn owl. J Comp Neurol 314(2):306–318

Carr CE, Boudreau RE (1993) An axon with a myelinated initial segment in the bird auditory system. Brain Res 628(1–2):330–334

Cherubini E, Gaiarsa JL, Ben-Ari Y (1991) GABA: an excitatory transmitter in early postnatal life. Trends Neurosci 14:515–519

Chuhma N, Ohmori H (1998) Postnatal development of phase-locked high-fidelity synaptic transmission in the medial nucleus of the trapezoid body of the rat. J Neurosci 18(1):512–520

Clark BA, Monsivais P, Branco T, London M, Hausser M (2005) The site of action potential initiation in cerebellar Purkinje neurons. Nat Neurosci 8:137–139

Code RA, Burd GD, Rubel EW (1989) Development of GABA immunoreactivity in brainstem auditory nuclei of the chick: ontogeny of gradients in terminal staining. J Comp Neurol 284(4):504–518

Conn PJ, Pin JP (1997) Pharmacology and functions of metabotropic glutamate receptors. Annu Rev Pharmacol Toxicol 37:205–237

Cook DL, Schwindt PC, Grande LA, Spain WJ (2003) Synaptic depression in the localization of sound. Nature 421(6918):66–70

Coombs JS, Curtis DR, Eccles JC (1957) The generation of impulses in motoneurones. J Physiol 139(2):232–249

Cudmore RH, Fronzaroli-Molinieres L, Giraud P, Debanne D (2010) Spike-time precision and network synchrony are controlled by the homeostatic regulation of the D-type potassium current. J Neurosci 30(38):12885–12895

Dasika VK, White JA, Carney LH, Colburn HS (2005) Effects of inhibitory feedback in a network model of avian brain stem. J Neurophysiol 94(1):400–414

DiFrancesco D, Tromba C (1988) Muscarinic control of the hyperpolarization-activated current (if) in rabbit sino-atrial node myocytes. J Physiol 405:493–510

DiFrancesco D, Ferroni A, Mazzanti M, Tromba C (1986) Properties of the hyperpolarizing-activated current (if) in cells isolated from the rabbit sino-atrial node. J Physiol 377:61–88

Eyre MD, Renzi M, Farrant M, Nusser Z (2012) Setting the time course of inhibitory synaptic currents by mixing multiple GABA(A) receptor α subunit isoforms. J Neurosci 32(17):5853–5867

Farrant M, Nusser Z (2005) Variations on an inhibitory theme: phasic and tonic activation of GABA(A) receptors. Nat Rev Neurosci 6(3):215–229

Fitzpatrick DC, Batra R, Stanford TR, Kuwada S (1997) A neuronal population code for sound localization. Nature 388(6645):871–874

Franken TP, Roberts MT, Wei L, Golding NL, Joris PX (2015) In vivo coincidence detection in mammalian sound localization generates phase delays. Nat Neurosci 18(3):444–452

Fujita I, Konishi M (1991) The role of GABAergic inhibition in processing of interaural time difference in the owl's auditory system. J Neurosci 11(3):722–739
Fukui I, Sato T, Ohmori H (2006) Improvement of phase information at low sound frequency in nucleus magnocellularis of the chicken. J Neurophysiol 6(2):633–641
Fukui I, Ohmori H (2003) Developmental changes in membrane excitability and morphology of neurons in the nucleus angularis of the chicken. J Physiol 548(Pt 1):219–232
Fukui I, Ohmori H (2004) Tonotopic gradients of membrane and synaptic properties for neurons of the chicken nucleus magnocellularis. J Neurosci 24(34):7514–7523
Funabiki K, Koyano K, Ohmori H (1998) The role of GABAergic inputs for coincidence detection in the neurones of nucleus laminaris of the chick. J Physiol 508(Pt 3):851–869
Tang ZQ, Gao H, Lu Y (2009a) Control of a depolarizing GABAergic input in an auditory coincidence detection circuit. J Neurophysiol 102(3):1672–1683
Gardner SM, Trussell LO, Oertel D (2001) Correlation of AMPA receptor subunit composition with synaptic input in the mammalian cochlear nuclei. J Neurosci 21(18):7428–7437
Gilman AG (1987) G proteins: transducers of receptor-generated signals. Annu Rev Biochem 56:615–649
Goldberg JM, Brown PB (1969) Response of binaural neurons of dog superior olivary complex to dichotic tonal stimuli: some physiological mechanisms of sound localization. J Neurophysiol 32(4):613–636
Goldstein PA, Elsen FP, Ying SW, Ferguson C, Homanics GE, Harrison NL (2002) Prolongation of hippocampal miniature inhibitory postsynaptic currents in mice lacking GABAA receptor 1 subunit. J Neurophysiol 88:3208–3217
Grissmer S, Nguyen AN, Aiyar J, Hanson DC, Mather RJ, Gutman GA, Karmilowicz MJ, Auperin DD, Chandy KG (1994) Pharmacological characterization of five cloned voltage-gated K+ channels, types Kv1.1, 1.2, 1.3, 1.5, and 3.1, stably expressed in mammalian cell lines. Mol Pharmacol 45(6):1227–1234
Grothe B, Sanes DH (1994) Synaptic inhibition influences the temporal coding properties of medial superior olivary neurons: an in vitro study. J Neurosci 14(3 Pt 2):1701–1709
Grothe B (2003) New roles for synaptic inhibition in sound localization. Nat Rev Neurosci 4(7):540–550
Grubb MS, Burrone J (2010) Activity-dependent relocation of the axon initial segment fine-tunes neuronal excitability. Nature 465(7301):1070–1074
Hackett JT, Jackson H, Rubel EW (1982) Synaptic excitation of the second and third order auditory neurons in the avian brain stem. Neuroscience 7:1455–1469
Harper NS, McAlpine D (2004) Optimal neural population coding of an auditory spatial cue. Nature 430(7000):682–686
Harvey AL (2001) Twenty years of dendrotoxins. Toxicon 39(1):15–26
van der Heijden M, Lorteije JA, Plauška A, Roberts MT, Golding NL, Borst JG (2013) Directional hearing by linear summation of binaural inputs at the medial superior olive. Neuron 78(5):936–948
Hopkins WF, Allen ML, Houamed KM, Tempel BL (1994) Properties of voltage-gated K currents expressed in Xenopus oocytes by mKv1.1, mKv1.2 and their heteromultimers as revealed by mutagenesis of the dendrotoxin-binding site in mKv1.1. Pflügers Arch 428:382–390
Hossain WA, Antic SD, Yang Y, Rasband MN, Morest DK (2005) Where is the spike generator of the cochlear nerve? Voltage-gated sodium channels in the mouse cochlea. J Neurosci 25(29):6857–6868
Hyson RL, Overholt EM, Lippe WR (1994) Cochlear microphonic measurements of interaural time differences in the chick. Hear Res 81(1–2):109–118
Ishii TM, Takano M, Xie LH, Noma A, Ohmori H (1999) Molecular characterization of the hyperpolarization-activated cation channel in rabbit heart sinoatrial node. J Biol Chem 274(18):12835–12839
Iwasaki S, Takahashi T (2001) Developmental regulation of transmitter release at the calyx of held in rat auditory brainstem. J Physiol 534(Pt 3):861–871

Iwasaki S, Takahashi T (1998) Developmental changes in calcium channel types mediating synaptic transmission in rat auditory brainstem. J Physiol 509(Pt 2):419–423
Jeffress LA (1948) A place theory of sound localization. J Comp Physiol Psychol 41:35–39
Jhaveri S, Morest DK (1982) Sequential alterations of neuronal architecture in nucleus magnocellularis of the developing chicken: a Golgi study. Neuroscience 7(4):837–853
Joris PX, Carney LH, Smith PH, Yin TC (1994) Enhancement of neural synchronization in the anteroventral cochlear nucleus. I. Responses to tones at the characteristic frequency. J Neurophysiol 71:1022–1036
Joris P, Yin TC (2007) A matter of time: internal delays in binaural processing. Trends Neurosci 30(2):70–78
Joseph AW, Hyson RL (1993) Coincidence detection by binaural neurons in the chick brain stem. J Neurophysiol 69(4):1197–1211
Kew JN, Kemp JA (2005) Ionotropic and metabotropic glutamate receptor structure and pharmacology. Psychopharmacology 79(1):4–29
Khaliq ZM, Raman IM (2006) Relative contributions of axonal and somatic Na channels to action potential initiation in cerebellar Purkinje neurons. J Neurosci 26(7):1935–1944
Klump GM (2000) Sound localization in birds. In: Dooling RJ, Fay RR, Popper AN (eds) Comparative hearing: birds and reptiles. Springer, New York, pp 249–307
Knudsen EI, Blasdel GG, Konishi M (1979) Sound localization by the barn owl (Tyto alba) measured with the search coil technique. J Comp Physiol 133:1–11
Köppl C, Carr CE (2008) Maps of interaural time difference in the chicken's brainstem nucleus laminaris. Biol Cybern 98(6):541–559
Köppl C, Carr CE (1997) Low-frequency pathway in the barn owl's auditory brainstem. J Comp Neurol 378(2):265–282
Köppl C (1994) Auditory nerve terminals in the cochlear nucleus magnocellularis: differences between low and high frequencies. J Comp Neurol 339(3):438–446
Köppl C (1997) Phase locking to high frequencies in the auditory nerve and cochlear nucleus magnocellularis of the barn owl, Tyto alba. J Neurosci 17(9):3312–3321
Kole MH, Ilschner SU, Kampa BM, Williams SR, Ruben PC, Stuart GJ (2008) Action potential generation requires a high sodium channel density in the axon initial segment. Nat Neurosci 11(2):178–186
Kole MH, Stuart GJ (2008) Is action potential threshold lowest in the axon? Nat Neurosci 11(11):1253–1255
Koyano K, Funabiki K, Ohmori H (1996) Voltage-gated ionic currents and their roles in timing coding in auditory neurons of the nucleus magnocellularis of the chick. Neurosci Res 26(1):29–45
Kuba H, Koyano K, Ohmori H (2002a) Development of membrane conductance improves coincidence detection in the nucleus laminaris of the chicken. J Physiol 540(Pt 2):529–542
Kuba H, Koyano K, Ohmori H (2002b) Synaptic depression improves coincidence detection in the nucleus laminaris in brainstem slices of the chick embryo. Eur J Neurosci 15(6):984–990
Kuba H, Yamada R, Ohmori H (2003) Evaluation of the limiting acuity of coincidence detection in nucleus laminaris of the chicken. J Physiol 552(Pt 2):611–620
Kuba H, Ishii TM, Ohmori H (2006) Axonal site of spike initiation enhances auditory coincidence detection. Nature 444(7122):1069–1072
Kuba H, Yamada R, Fukui I, Ohmori H (2005) Tonotopic specialization of auditory coincidence detection in nucleus laminaris of the chick. J Neurosci 25(8):1924–1934
Kuba H, Ohmori H (2009) Roles of axonal sodium channels in precise auditory time coding at nucleus magnocellularis of the chick. J Physiol 587(1):87–100
Kuba H, Oichi Y, Ohmori H (2010) Presynaptic activity regulates Na(+) channel distribution at the axon initial segment. Nature 465(7301):1075–1078
Kuba H, Adachi R, Ohmori H (2014) Activity-dependent and activity-independent development of the axon initial segment. J Neurosci 34(9):3443–3453
Lachica EA, Rübsamen R, Zirpel L, Rubel EW (1995) Glutamatergic inhibition of voltage-operated calcium channels in the avian cochlear nucleus. J Neurosci 15(3):1724–1734

Lachica EA, Rübsamen R, Rubel EW (1994) GABAergic terminals in nucleus magnocellularis and laminaris originate from the superior olivary nucleus. J Comp Neurol 348(3):403–418
Larsen ON, Dooling RJ, Michelsen A (2006) The role of pressure difference reception in the directional hearing of budgerigars (Melopsittacus undulatus). J Comp Physiol A Neuroethol Sens Neural Behav Physiol 192:1063–1072
Leão RM, Von Gersdorff H (2002) Noradrenaline increases high-frequency firing at the calyx of held synapse during development by inhibiting glutamate release. J Neurophysiol 87(5):2297–2306
Levin MD, Kubke MF, Schneider M, Wenthold R, Carr CE (1997) Localization of AMPA-selective glutamate receptors in the auditory brainstem of the barn owl. J Comp Neurol 378(2):239–253
Lippe W, Rubel EW (1985) Ontogeny of tonotopic organization of brain stem auditory nuclei in the chicken: implications for development of the place principle. J Comp Neurol 237(2):273–289
Ludwig A, Zong X, Jeglitsch M, Hofmann F, Biel M (1998) A family of hyperpolarization-activated mammalian cation channels. Nature 393:587–591
Ludwig A, Zong X, Stieber J, Hullin R, Hofmann F, Biel M (1999) Two pacemaker channels from human heart with profoundly different activation kinetics. EMBO J 18(9):2323–2329
Maki K, Furukawa S (2005) Acoustical cues for sound localization by the Mongolian gerbil, Meriones unguiculatus. J Acoust Soc Am 118(2):872–886
Manley GA, Yates GK, Köppl C (1988) Auditory peripheral tuning: evidence for a simple resonance phenomenon in the lizard Tiliqua. Hear Res 33(2):181–189
Manley GA, Kaiser A, Brix J, Gleich O (1991) Activity patterns of primary auditory-nerve fibres in chickens: development of fundamental properties. Hear Res 57(1):1–15
McAlpine D, Jiang D, Palmer AR (2001) A neural code for low-frequency sound localization in mammals. Nat Neurosci 4(4):396–401
McClellan AM, Twyman RE (1999) Receptor system response kinetics reveal functional subtypes of native murine and recombinant human GABAA receptors. J Physiol 515(Pt 3):711–727
Midorikawa M, Okamoto Y, Sakaba T (2014) Developmental changes in Ca^{2+} channel subtypes regulating endocytosis at the calyx of held. J Physiol 592(16):3495–3510
Mody I, Pearce RA (2004) Diversity of inhibitory neurotransmission through GABA(A) receptors. Trends Neurosci 27(9):569–575
Mogdans J, Knudsen EI (1994) Representation of interaural level difference in the VLVp, the first site of binaural comparison in the barn owl's auditory system. Hear Res 74(1–2):148–164
Moiseff A, Konishi M (1981) Neuronal and behavioral sensitivity to binaural time differences in the owl. J Neurosci 1(1):40–48
Moons L, D'Hondt E, Pijcke K, Vandesande F (1995) Noradrenergic system in the chicken brain: immunocytochemical study with antibodies to noradrenaline and dopamine-beta-hydroxylase. J Comp Neurol 360(2):331–348
Mosbacher J, Schoepfer R, Monyer H, Burnashev N, Seeburg PH, Ruppersberg JP (1994) A molecular determinant for submillisecond desensitization in glutamate receptors. Science 266(5187):1059–1062
Nishino E, Yamada R, Kuba H, Hioki H, Furuta T, Kaneko T, Ohmori H (2008) Sound-intensity-dependent compensation for the small interaural time difference cue for sound source localization. J Neurosci 28(28):7153–7164
Okuda H, Yamada R, Kuba H, Ohmori H (2013) Activation of metabotropic glutamate receptors improves the accuracy of coincidence detection by presynaptic mechanisms in the nucleus laminaris of the chick. J Physiol 591(1):365–378
Olsen JF, Knudsen EI, Esterly SD (1989) Neural maps of interaural time and intensity differences in the optic tectum of the barn owl. J Neurosci 9(7):2591–2605
Otis TS, Trussell LO (1996) Inhibition of transmitter release shortens the duration of the excitatory synaptic current at a calyceal synapse. J Neurophysiol 76(5):3584–3588
Otis TS, Raman IM, Trussell LO (1995) AMPA receptors with high Ca2+ permeability mediate synaptic transmission in the avian auditory pathway. J Physiol 482(Pt 2):309–315
Overholt EM, Rubel EW, Hyson RL (1992) A circuit for coding interaural time differences in the chick brainstem. J Neurosci 12(5):1698–1708

Palmer LM, Stuart GJ (2006) Site of action potential initiation in layer 5 pyramidal neurons. J Neurosci 26(6):1854–1863
Pape HC (1996) Queer current and pacemaker: the hyperpolarization-activated cation current in neurons. Annu Rev Physiol 58:299–327
Parameshwaran S, Carr CE, Perney TM (2001) Expression of the Kv3.1 potassium channel in the avian auditory brainstem. J Neurosci 21(2):485–494
Rubel EW, Parks TN (1975) Organization and development of brain stem auditory nuclei of the chicken: tonotopic organization of n. magnocellularis and n. laminaris. J Comp Neurol 164(4):411–433
Parks TN, Rubel EW (1978) Organization and development of the brain stem auditory nuclei of the chicken: primary afferent projections. J Comp Neurol 180(3):439–448
Pecka M, Brand A, Behrend O, Grothe B (2008) Interaural time difference processing in the mammalian medial superior olive: the role of glycinergic inhibition. J Neurosci 28(27):6914–6925
Peña JL, Viete S, Albeck Y, Konishi M (1996) Tolerance to sound intensity of binaural coincidence detection in the nucleus laminaris of the owl. J Neurosci 16(21):7046–7054
Raman IM, Zhang S, Trussell LO (1994) Pathway-specific variants of AMPA receptors and their contribution to neuronal signaling. J Neurosci 14(8):4998–5010
Raman IM, Trussell LO (1992) The kinetics of the response to glutamate and kainate in neurons of the avian cochlear nucleus. Neuron 9(1):173–186
Rasband MN (2010) Composition, assembly, and maintenance of excitable membrane domains in myelinated axons. Semin Cell Dev Biol 22(2):178–184
Rathouz M, Trussell L (1998) Characterization of outward currents in neurons of the avian nucleus magnocellularis. J Neurophysiol 80(6):2824–2835
Ravindranathan A, Donevan SD, Sugden SG, Greig A, Rao MS, Parks TN (2000) Contrasting molecular composition and channel properties of AMPA receptors on chick auditory and brainstem motor neurons. J Physiol 523(Pt 3):667–684
Reyes AD, Rubel EW, Spain WJ (1994) Membrane properties underlying the firing of neurons in the avian cochlear nucleus. J Neurosci 14(9):5352–5364
Roberts MT, Seeman SC, Golding NL (2013) A mechanistic understanding of the role of feedforward inhibition in the mammalian sound localization circuitry. Neuron 78(5):923–935
Ryals BM, Rubel EW (1982) Patterns of hair cell loss in chick basilar papilla after intense auditory stimulation. Frequency organization. Acta Otolaryngol 93(1–6):205–210
Salvi RJ, Saunders SS, Powers NL, Boettcher FA (1999) Discharge patterns of cochlear ganglion neurons in the chicken. J Comp Physiol A 170:227–241
Santoro B, Tibbs GR (1999) The HCN gene family: molecular basis of the hyperpolarization-activated pacemaker channels. Ann N Y Acad Sci 868:741–764
Santoro B, Liu DT, Yao H, Bartsch D, Kandel ER, Siegelbaum SA, Tibbs GR (1998) Identification of a gene encoding a hyperpolarization-activated pacemaker channel of brain. Cell 93:717–729
Sato T, Fukui I, Ohmori H (2010) Interaural phase difference modulates the neural activity in the nucleus angularis and improves the processing of level difference cue in the lateral lemniscal nucleus in the chicken. Neurosci Res 66(2):198–212
Saunders JC, Ventetuolo CE, Plontke SK, Weiss BA (2002) Coding of sound intensity in the chick cochlear nerve. J Neurophysiol 88(6):2887–2898
Schroeder CI, Lewis RJ, Adams DJ (2013) Block of VoltageGated Calcium Channels by Peptide Toxins. in Madame Curie Bioscience database – NCBI Bookshelf: Landes Bioscience
Seidl AH, Rubel EW, Harris DM (2010) Mechanisms for adjusting interaural time differences to achieve binaural coincidence detection. J Neurosci 30(1):70–80
Seidl AH, Rubel EW, Barría A (2014) Differential conduction velocity regulation in ipsilateral and contralateral collaterals innervating brainstem coincidence detector neurons. J Neurosci 34(14):4914–4919
Seifert R, Scholten A, Gauss R, Mincheva A, Lichter P, Kaupp UB (1999) Molecular characterization of a slowly gating human hyperpolarization-activated channel predominantly expressed in thalamus, heart, and testis. Proc Natl Acad Sci U S A 96(16):9391–9396

Shibata S, Kakazu Y, Okabe A, Fukuda A, Nabekura J (2004) Experience-dependent changes in intracellular Cl– regulation in developing auditory neurons. Neurosci Res 48:211–220
Smith PH, Joris PX, Carney LH, Yin TC (1991) Projections of physiologically characterized globular bushy cell axons from the cochlear nucleus of the cat. J Comp Neurol 304(3):387–407
Smith PH, Joris PX, Yin TC (1998) Anatomy and physiology of principal cells of the medial nucleus of the trapezoid body (MNTB) of the cat. J Neurophysiol 79(6):3127–3142
Smith DJ, Rubel EW (1979) Organization and development of brain stem auditory nuclei of the chicken: dendritic gradients in nucleus laminaris. J Comp Neurol 186(2):213–239
Spitzer MW, Semple MN (1995) Neurons sensitive to interaural phase disparity in gerbil superior olive: diverse monaural and temporal response properties. J Neurophysiol 73(4):1668–1690
Stuart G, Spruston N, Sakmann B, Häusser M (1997) Action potential initiation and backpropagation in neurons of the mammalian CNS. Trends Neurosci 20(3):125–131
Sullivan WE, Konishi M (1984) Segregation of stimulus phase and intensity coding in the cochlear nucleus of the barn owl. J Neurosci 4(7):1787–1799
Takahashi TT, Konishi M (1988) Projections of nucleus angularis and nucleus laminaris to the lateral lemniscal nuclear complex of the barn owl. J Comp Neurol 274(2):212–238
Takahashi T, Forsythe ID, Tsujimoto T, Barnes-Davies M, Onodera K (1996) Presynaptic calcium current modulation by a metabotropic glutamate receptor. Science 274(5287):594–597
Takahashi T, Kajikawa Y, Tsujimoto T (1998) G-Protein-coupled modulation of presynaptic calcium currents and transmitter release by a GABAB receptor. J Neurosci 18(9):3138–3146
Takahashi T, Moiseff A, Konishi M (1984) Time and intensity cues are processed independently in the auditory system of the owl. J Neurosci 4(7):1781–1786
Tang ZQ, Gao H, Lu Y (2009b) Control of a depolarizing GABAergic input in an auditory coincidence detection circuit. J Neurophysiol 102(3):1672–1683
Tilney LG, Saunders JC (1983) Actin filaments, stereocilia, and hair cells of the bird cochlea. I. Length, number, width, and distribution of stereocilia of each hair cell are related to the position of the hair cell on the cochlea. J Cell Biol 96(3):807–821
Tollin DJ, Yin TC (2005) Interaural phase and level difference sensitivity in low-frequency neurons in the lateral superior olive. J Neurosci 25(46):10648–10657
Trussell LO (1997) Cellular mechanisms for preservation of timing in central auditory pathways. Curr Opin Neurobiol 7(4):487–492
Trussell LO (1999) Synaptic mechanisms for coding timing in auditory neurons. Annu Rev Physiol 61:477–496
Vicini S, Ferguson C, Prybylowski K, Kralic J, Morrow AL, Homanics GE (2001) GABA(A) receptor alpha1 subunit deletion prevents developmental changes of inhibitory synaptic currents in cerebellar neurons. J Neurosci 21(9):3009–3016
von Bartheld CS, Code RA, Rubel EW (1989) GABAergic neurons in brainstem auditory nuclei of the chick: distribution, morphology, and connectivity. J Comp Neurol 287(4):470–483
von Gersdorff H, Borst JG (2002) Short-term plasticity at the calyx of held. Nat Rev Neurosci 3(1):53–64
Wang LY, Gan L, Forsythe ID, Kaczmarek LK (1998) Contribution of the Kv3.1 potassium channel to high-frequency firing in mouse auditory neurones. J Physiol 509(Pt 1):183–194
Wang HL, Zhang Z, Hintze M, Chen L (2011) Decrease in calcium concentration triggers neuronal retinoic acid synthesis during homeostatic synaptic plasticity. J Neurosci 31(49):17764–17771
Warchol ME, Dallos P (1990) Neural coding in the chick cochlear nucleus. J Comp Physiol A 166(5):721–734
Watanabe M, Fukuda A (2015) Development and regulation of chloride homeostasis in the central nervous system. Front Cell Neurosci 9:371
Xu-Friedman MA, Regehr WG (2005) Dynamic-clamp analysis of the effects of convergence on spike timing. I. Many synaptic inputs. J Neurophysiol 94(4):2512–2525
Yamada R, Kuba H, Ishii TM, Ohmori H (2005) Hyperpolarization-activated cyclic nucleotide-gated cation channels regulate auditory coincidence detection in nucleus laminaris of the chick. J Neurosci 25(39):8867–8877

Yamada R, Okuda H, Kuba H, Nishino E, Ishii TM, Ohmori H (2013) The cooperation of sustained and phasic inhibitions increases the contrast of ITD-tuning in low-frequency neurons of the chick nucleus laminaris. J Neurosci 33(9):3927–3938

Yang L, Monsivais P, Rubel EW (1999) The superior olivary nucleus and its influence on nucleus laminaris: a source of inhibitory feedback for coincidence detection in the avian auditory brainstem. J Neurosci 19:2313–2325

Yin TC, Chan JC (1990) Interaural time sensitivity in medial superior olive of cat. J Neurophysiol 64(2):465–488

Young SR, Rubel EW (1983) Frequency-specific projections of individual neurons in chick brainstem auditory nuclei. J Neurosci 3(7):1373–1378

Zhang S, Trussell LO (1994) A characterization of excitatory postsynaptic potentials in the avian nucleus magnocellularis. J Neurophysiol 72(2):705–718

Chapter 3
Central Auditory Processing

Auditory sensation does not mean only hearing a sound but is also a basis of our language communication and logical thinking. As described in the preceding chapters, auditory signals are extensively processed in the ascending pathway, and specific acoustic signals are extracted, integrated and encoded into a more representative information before arriving to the telencephalon, such as ITD (interaural time difference) in the NL or in the MSO and ILD (interaural level difference) in the LL or in the LSO, in birds or in mammals, respectively. In the telencephalon, we believe that auditory signals are integrated with various other sensory signals and even with motor information. Then, the auditory signal is projected back to the subcortical nuclei through the corticofugal projection, and acoustic sensation will be refined and reformed as activity of neurons. These feedback controls will update the auditory sensation continuously.

In early developmental days, animals have a critical period to form a neural representation of sensation. During the critical period, the auditory space map in the inferior colliculus (IC) is arranged under the control of visual sensation in the barn owl (Hyde and Knudsen 2002). After the end or closure of the critical period, the neural activity is stabilized. Surprisingly, even after the closure of critical period in adulthood, the cortical representation of tonotopy can be rearranged under a special condition of environment in the adult rats, by exposure to chronic noise (Zhou et al. 2011). That is, chronic exposure to a moderate level noise has an effect to reopen the critical period of auditory sensation. Accordingly, even the adult auditory cortex maintains its plasticity (Weinberger 1995; Buonomano and Merzenich 1998; Froemke et al. 2007). Other examples of plasticity of the adult auditory cortex could be found in the song learning of songbirds (Brainard and Doupe 2002), where student birds listen to the tutor's song and master the species-specific song by practice. In the adult bat, frequency tuning and other features are fine-tuned to increase or decrease the responsibility in neurons of IC and auditory cortex depending on the degree of matching of features between these neurons (Suga and Ma 2003). However, exact neural mechanisms that underlie these auditory cortical plasticities are not yet resolved, although each phenomenon is well described in various species.

H. Ohmori, *Auditory Information Processing*,
https://doi.org/10.1007/978-981-32-9713-5_3

I will review in this chapter several topics on neural plasticity in the auditory cortex and in IC.

At the end of this chapter, I will describe two emerging technologies from my laboratory. Those technologies will provide some advancement in our understanding of the brain function, even in small steps. One is a tracing technology of neural projections across several synapses for a long distance in the chicken, using A3V (avian adeno-associated virus). Understanding of exact neural connections is the basis to elucidate brain function. The other is a development of photometric patch electrode (PME) recording system that can record both electrical activity and optical signals simultaneously from neurons at any depth of brain structure. Neural activity accompanies intracellular metabolic processes in order to maintain the activity itself or to consolidate the adaptive outcome of the activity. Activity of cellular metabolism can be reported by optical signal such as the Ca^{2+} fluorescence signal. Accordingly, simultaneous measurement of electrical and optical signals of neurons could be one of the ideal methods to elucidate higher brain functions.

3.1 Learning Songs of Songbirds

Many experiments have been conducted to explore the mechanism how nervous system works, by using variety of animal species that have specialized features. One typical animal is the squid. Giant axon of the squid advanced our understanding on the mechanisms of membrane excitation (Hodgkin et al. 1952), and the giant synapse of squid facilitated the understanding of synaptic transmission (Bullock and Hagiwara 1957). Songbirds are one of those specialist animals that have neural circuits specialized for learning songs and practicing vocalization. Motor circuitry of vocalization is gradually shaped to reproduce the learned songs by practice or performance-based feedback. There are several sensory processes for the song learning. First is the recognition of template song or a step of exposure to a tutor song. Second is the activity of some genetically programmed circuitry that recognizes and learns or recalls the species-specific song. Third is the step of memorization and practice of the song. The final step is to complete the song or to crystallize the phase of song development. Exposure to a tutor song is discussed to have particular importance to influence the activity of neurons and the plasticity within HVC (the hyperstriatum ventrale, pars caudale), which is a nucleus crucial for song imitation and learning (Roberts et al. 2010; Vallentin et al. 2016; Roberts et al. 2012).

Auditory signals including the song are first fed in to field L in the telencephalon. Field L is the primary auditory area of birds and is indispensable for generation of songs and vocal communication (Brainard and Doupe 2002; Elliott and Theunissen 2011). Field L is a large region in the caudal telencephalon and is a highly integrative center for intracortical auditory signal processing (Fig. 3.1). Field L is made of three major regions L1, L2, and L3 (Wild et al. 1993). Field L2 receives inputs from the thalamus (nucleus ovoidalis, nOV), so that field L2 is considered homologous to the layer IV of mammalian auditory cortex (Wang et al. 2010). Neurons in field L and the secondary auditory area, the caudal mesopallium (CM; Vu et al. 1994), respond robustly to auditory stimuli such as white noise, the bird's own song, and songs of the same species. Furthermore, neurons in field L are sources of auditory signals to the nucleus interfacialis of the nidopallium (NIf) (Keller and Hahnloser 2009). NIf is necessary for learning and forms a part of sensory-motor loop that includes HVC.

The nucleus HVC acts as a sensory-motor interface and generates commands essential for the song production (Nottebohm et al. 1982; Bauer et al. 2008; Akutagawa and Konishi 2010). HVC projects to RA (robust nucleus of the archistriatum) in two routes; one is directly to RA and has a function for song motor pathway (Fig. 3.1). The other is through the projection to AFP (anterior forebrain pathway). AFP is interconnected complex of nuclei connecting Area X, DLM, LMAN, and RA and is known critical for the song learning (Bottjer and Johnson 1997; Vu et al. 1994; Boettiger and Doupe 2001; Aronov et al. 2008).

The direct motor pathway from HVC to RA is necessary for normal song production. RA controls the bird's vocal organ and projects to the trachea-syringeal portion of the hypoglossal nucleus (Fig. 3.1). The hypoglossal nucleus is involved in control of respiration during singing (Brainard and Doupe 2002). AFP is connected to the motor nuclei HVC both at the Area X as inputs and RA as outputs.

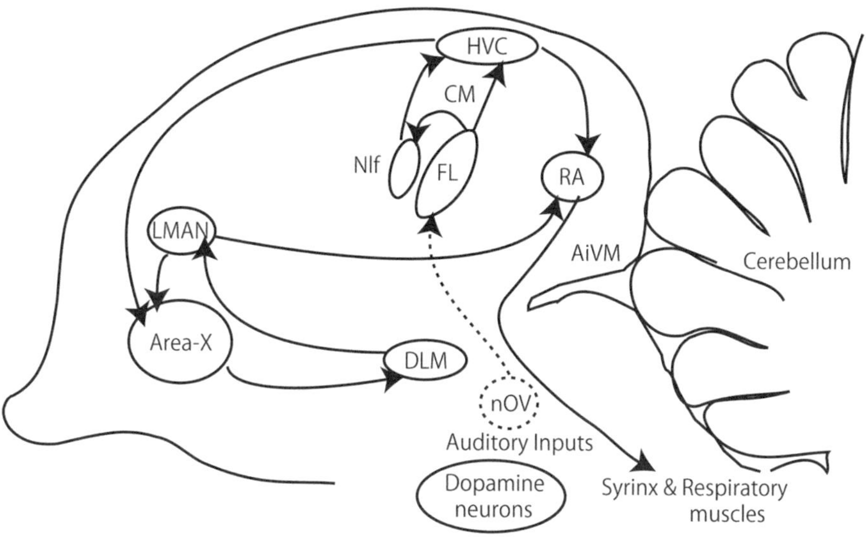

Fig. 3.1 Schemaic representation of the circuit for song learning and vocalization
Solid lines illustrate connections of nuclei in the song circuits and the arrow head indicates the direction of control. Broken line with an arrow head indicates the sound input projection from nOV to field L. *nOV* nucleus ovoidalis, *FL* field L, *Area X* Area X of the medial striatum, *DLM* medial dorsolateral nucleus of the thalamus, *LMAN* lateral magnocellular nucleus of the anterior neostriatum, *RA* robust nucleus of the archistriatum, *NIF* nucleus interfacialis of the nidopallium, *HVC* hyperstriatum ventrale pars caudale, *AiVM* ventromedial nucleus of the intermediate archistriatum

Neurons in AFP are tuned to the bird's own song and tutor song. This dual tuning of songs should be useful for the comparison of songs and is essential for song learning. Lesions of AFP in juveniles severely disrupt song learning.

The AFP circuit activity is highly enhanced in the sensory critical period or in early sensorimotor learning phases (Iyengar et al. 1999; Boettiger and Doupe 2001). Volumes of these nuclei grew in juvenile male, but in the adulthood volumes decreased. Projection from LMAN to RA undergoes substantial refinement during early stages of song learning (Iyengar et al. 1999). A long-term synaptic plasticity of both LTP (long-term potentiation) and LTD (long-term depression) types is required for song learning in LMAN (Boettiger and Doupe 2001). Blockade of NMDA receptors in LMAN during learning a tutor song prevented birds from producing a good copy of the tutor song (Basham et al. 1996), suggesting a presence of NMDA receptors-dependent long-term plasticity in LMAN during sensory learning. It is also known that NMDA receptors expression in LMAN was downregulated significantly by the end of sensory critical period (Aamodt et al. 1992). Consistently, lesions of LMAN during song acquisition period disturb song development, whereas similar lesioning in adult birds does not affect normal song production (Bottjer et al. 1984; Sohrabji et al. 1990; Scharff and Nottebohm 1991; Nordeen and Nordeen 1992; Morrison and Nottebohm 1993).

Learning and maintenance of the song requires auditory feedback by practicing, such as listening to and imitating the tutor song (Leonardo and Konishi 1999). Auditory feedback is controlled by a dopaminergic projection from VTA (ventral tegmental area, a midbrain dopamine region) to Area X (Bottjer and Johnson 1997; Luo Ding Perkel 2001; Doupe et al. 2004). A contribution of dopaminergic neurons in the midbrain periaqueductal gray (PAG) that project to HVC is also demonstrated recently to contribute to detect the presence of a tutor and to facilitate copying the tutor song in juvenile zebra finches (Tanaka et al. 2018). The projection from Area X to the dorsolateral thalamic nucleus DLM is GABAergic and inhibitory (Luo and Perkel 1999). The overall connection for the song control is a closed loop including LMAN-Area X-DLM-LMAN, which appears similar to the direct pathway of mammalian basal ganglia (Bottjer and Johnson 1997) (Fig. 3.1). In the mammalian basal ganglia, the overall connection is a closed loop including cortex-striatum-globus pallidus-thalamus-cortex. Here the dopaminergic inputs are made from substantia nigra and excite striatum neurons that inhibit globus pallidus. This results in the disinhibition of thalamic neurons to the cortex.

The sensitive or critical period affects critically this song system (Doupe et al. 2004). The timing of sensitive period depends on experience as well as hormones. Gonadal hormones are necessary to enhance the song learning and induce morphological changes in HVC and RA (Nottebohm 1980). These two nuclei in male canaries are larger in spring when they are producing stable adult song than in fall (Nottebohm 1981) and are larger in males that learn songs than in females that normally do not sing. The size increase of these nuclei was accompanied with increases of cell number and extension of dendritic arbors (Arnold et al. 1986; DeVoogd and Nottebohm 1981). A high-level testosterone was not required for the onset of song practice, but it was reported necessary for the song crystallization (Marler et al. 1988; Whaling et al. 1998).

All these observations provide important clues to the understanding of network building during critical period or seasonal learning period of birds' song. Hormonal effects and contributions of NMDA receptors are observed. The circuits contributing for song learnings are well described. However, we do not know yet the neuronal details of song system, such as how the individual song syllable is composed and stored in the neural pattern generator and how the sequence of syllables is formed to construct a bird's song. Detailed network activity of neural pattern generators should be resolved. We also do not know the details of sensory motor feedback mechanisms that shape and maintain the songs by listening to and by repeated practicing of the songs. It is extremely interesting to explore these neural networks that are responsible for the pattern generation of song syllables and for constructing a bird's songs.

3.2 Auditory Cortex Plasticity

The auditory cortex of adult rats can restore the capacity of plasticity when exposed for a long time to acoustic noise. This plastic state was again returned to the non-plastic state of adult normal cortex by returning the animal to natural acoustic environments for several weeks (Zhou et al. 2011). After the long-time noise exposure, neurons in the primary auditory cortex became less tuned to sound frequencies than neurons in control rats. Since ABR (auditory brainstem response) was normal even after the noise exposure, the noise exposure did not hurt the peripheral hearing capability.

A chronic exposure to moderate level acoustic noise (65 dB SPL white noise for 6 weeks beginning at 8th postnatal week) distorted the tonotopy of primary auditory cortex in juvenile or in adult rats. Normally the critical period of rats is closed about a month (4 weeks) after the birth, and the auditory cortex becomes non-plastic. Around that moment rats are approaching sexual maturity. Nevertheless, after the noise exposure, high-frequency area of the cortex was exaggerated, and low-frequency areas were reduced. Neurons in the primary auditory cortex became responsible less selectively to the sound frequency and failed to follow repetitive stimulus at high rate. That is, in normal rats neurons responded to 10 pulses per second (pps), but after the noise exposure rats could follow repetitive stimulus below 7 pps. Phase responsibility was also reduced to the stimulus of low-frequency repetition. Synchronized spontaneous firing activity that occurred within ±10 ms was reduced the incidence when evaluated by recordings with two electrodes from nearby neurons (Zhou et al. 2011). More interestingly, these distortions were returned to normal when animals were returned to the normal acoustic environments for 8 weeks after the noise exposure. Furthermore, exposure to a patterned sound stimulus (7 kHz tone for 1 week after the chronic noise exposure) enlarged the areas of primary auditory cortex that responded selectively to 7 kHz. The 7 kHz tone exposure was started immediately after the end of noise exposure epoch. These phenomena were equal to that occurred in infant rats during its critical period, and the phenomena were even observed in rats exposed to noise at 1-year-old, that is, well-matured animals. Accordingly, animals have expanded the cortical area that represented the trained frequency or the trained sound intensity, when they were trained by exposing to a specific frequency or a specific intensity sound (Polley et al. 2006).

The cortical plasticity is affected by balanced inputs of excitatory and inhibitory synapses. Noise exposure significantly reduced the expression of GABA receptors (alpha-1, beta-2/3 subunits) and NMDA receptors (NR2A and NR2B subunits) but there was no change in the expression level of AMPA receptors GluR2 subunit (Zhou et al. 2011). BDNF (brain-derived neurotrophic factor) level was low in the immature rats, and the noise exposure significantly reduced the expression of BDNF in primary auditory cortex of adult rats. Contribution for the critical period plasticity of $GABA_A$ receptors and NMDA receptors and BDNF level in various sensory modalities are repeatedly argued (Huang et al. 1999; Berardi et al. 2000; Fagiolini and Hensch 2000; Fagiolini et al. 2004). Reduced expressions of these substances in the auditory cortex of noise-exposed rats were completely returned to the normal level by returning the rats to a normal acoustic environment.

Reorganization of cortical tuning curves occurred after a short term acoustic stimulation paired with the electrical stimulation of the nucleus basalis (Froemke et al. 2007). The nucleus basalis is the main source of cortical acetylcholine (ACh). After the pairing, tone-evoked inhibitory postsynaptic currents were depressed in the primary auditory cortical neurons, while tone-evoked excitatory postsynaptic currents were potentiated. Reorganization of the tuning curve progressed for hours after the paired stimulation of nucleus basalis. The inhibition was gradually increased, in an activity dependent manner, and was slowly rebalanced with the persistently enhanced excitation. Since cortical application of atropine prevented the effects of paired stimulation to the nucleus basalis, the primary site of modulation was considered within the primary auditory cortex. ACh serves as a neuromodulator for receptive field plasticity. It is further discussed that the paired activity of nucleus basalis may be a synaptic correlate of attentiveness for novel or meaningful stimuli (Froemke et al. 2007).

These are amazing findings of adult cortical plasticity. Contribution of cholinergic activity was demonstrated in rebalancing of excitation and inhibition. It was demonstrated that small modulation on inhibitory inputs could greatly and rapidly affect the balanced operation of neural circuits based on the mechanisms of disinhibition. It is very interesting to explore further the mechanisms how the noise stimulus could destruct the balanced activity of neural circuits and result in the opening of once closed critical period.

3.3 Auditory Space Map in IC Is Plastic and Is Instructed by Visual Signals

During the critical period of barn owl postnatal development, visual system is important as an instructing signal to localize the auditory cues for the representation of space, the space map (Hyde and Knudsen 2002). The dominant auditory localization cues are the relative timing (ITDs, interaural time differences) and the relative level (ILDs, interaural level differences) of the sound detected at both ears. The central auditory system analyzes and associates these cues with locations in the space. This sound localization process is most extensively studied in the barn owl. The barn owl can hunt small animals in total darkness based only on the sound cues generated by moving small animals. ITD is the primary cue for localizing the sound source in the horizontal location, and ILD is the cue for the vertical location. As a consequence of asymmetric openings of two external ears in the barn owl, the left ear becomes more sensitive to the space below the horizontal plane of the animal and the right ear to the space above the horizontal plane (Knudsen and Konishi 1978). The sound spectrum is a cue to indicate the sound source location relative to the head and the ears, namely, in the front or the back of the head. Thus, the space map is formed with dimension of horizontal ITD axis and vertical ILD axis. The map is created in the lateral and anterior border of MLD (mesencephalicus lateralis dorsalis, avian inferior colliculus). The location corresponds to the external nucleus of inferior colliculus (ICX) in mammals and neurons there are sensitive to the high end of audible frequencies of the owl (4–8 kHz, Knudsen and Konishi 1979). The medial large portion of MLD (homologous to the core of inferior colliculus, ICC) is space independent, and neurons there are sensitive to sound frequency and are arranged in the order of frequency (tonotopy). Because these cues can vary over lifetime even in individuals as a result of development and aging, the auditory system needs to calibrate these localization cues continuously.

The reorientation of sound source localization cues was demonstrated by variety of experiments. One example is a monaurally occluded owl by plugging one ear. The owl had difficulty in localizing the sound source initially, but after several weeks the animal recovered the accuracy of orienting responses despite the presence of the earplug (Knudsen et al. 1984). Another example is the owl reared with prism glasses to redirect the vison. Animals wearing prism spectacles had a distorted vision and must learn to associate the auditory world and visual world into alignment. Over a period of many weeks, animals adjust their auditory orienting responses to the distorted optical cues. The auditory cue information in ICC is combined with that of ICX, and a new auditory space map is created. The new auditory space map is transferred to the optic tectum and contributes to adjust the auditory orienting behavior of the animal.

The tuning of neurons in ICC to auditory localization cues remains unchanged even by rearing with prism spectacles. However, in the ICX the localization cues were shaped by experience and were transformed into a new space map. Thus, the fiber projections from ICC to ICX were modified in the experience-dependent manner. Visual system instructs this re-direction activity of the auditory system. Creation of the new circuit is depended on the activity of NMDA receptors (Feldman and Knudsen 1994). In the early phase of prism learning, neurons in ICX responded to both normal and learned (or redirected) ITD. Application of AP5 (blocker of NMDA receptors) decreased the learned responses greatly in the ICX. Thus, NMDA receptors contributed to the functional expression of learned responses. The blocker of AMPA receptors (CNQX) blocked the normal responses more than the learned response. This differential contribution of glutamate receptors disappears over time when the visual field was returned to normal by removal of the prism spectacles.

$GABA_A$ receptors contribute importantly to the plasticity in ICX. The ITD tuning in the ICX was differentially inhibited by the activation of $GABA_A$ receptors during the prism spectacle learning. At the early phase of prism learning, the ITD tuning curve that was monitored in the optic tectum demonstrated a broad peak. The broadness was because of a presence of both normal ITD and adaptively learned ITD in the tuning curve. At the beginning of learning, the adaptive new responses were weak, but at the end of learning, the adaptive response became the major response. Surprisingly, even after the adaptive learning, the circuits corresponding to the normal response coexisted with the newly formed circuits, but the responses of normal circuits were inhibited by GABAergic activity (Zheng and Knudsen 1999). Suppression of the GABAergic inhibition brought up the normal ITD response again. Accordingly, the modulation of inhibition pattern by GABAergic activity has contributed critically to the process of adaptive plasticity of ITD in the ICX. The change of inhibition occurred in synchrony with the change in the excitation that occurred through activation of NMDA receptors. The visual activity in the optic tectum is proposed to encode the mismatch between the spatial representations of sound source and vision. Excitation of ICX neurons by the visual activity adjusts the representation of auditory localization cues in the ICX to match the visual representation of space (Brainard and Knudsen 1998; Knudsen 2002). It is extremely interesting to investigate how these mismatches between auditory and visual signals are detected in the optic tectum and are encoded as new signals that lead to the creation of new projections from ICC to ICX. It is furthermore interesting to investigate the neural mechanism that controls the differential modulation of GABAergic activity in the adaptive ITD plasticity.

3.4 Roles of Corticofugal Projection in the Plasticity of Auditory System

IC is considered as the obligatory relay center in the midbrain for ascending signals to the auditory cortex. Moreover, IC is an important target of corticofugal projection and is a site of interaction between the ascending and descending auditory signals. The corticofugal projection even modulates the cochlear function (Galambos 1956; Furukawa 1981) and likely makes the auditory periphery a target of auditory attention. A lack of corticofugal auditory feedback produces a profound deterioration of speech, as it happens if children became deaf early or even late in childhood. Cortico-collicular projection is the IC part of corticofugal projections and is predominantly ipsilateral. However, it is reported that a small percentage (10–15%) of cortico-collicular projection terminates in the contralateral IC in mammals (bat-Suga and Ma 2003; rat-Saldaña et al. 1996; cat- Winer et al. 1998; guinea pig-Schofield 2009).

The auditory signal processing was tuned and adjusted through the activity of cortico-collicular projection to be matched between IC and the auditory cortex (Yan and Suga 1998; Ma and Suga 2001, Jen and Zhou 2003, Zhou and Jen 2005; Yan and Ehret 2002, Yan et al. 2005). When electrical stimulation was applied to neurons in the auditory cortex, the response of IC neurons was tuned to match the features of stimulated neurons in the auditory cortex. For example, the frequency tuning was sharpened when the best frequency (BF) was matched between the stimulated auditory cortical neuron and the recorded IC neuron. When BF was not matched, the response of the IC neuron was inhibited at its BF, while the responsiveness at other frequencies was augmented. Accordingly, the cortico-collicular projection improved and reorganized the subcortical signal processing. Furthermore, the cortical modulation of IC activity is returned to cortical neurons as the ascending signal and modulates the signal processing of cortical neurons that have corresponding BF (Suga and Ma 2003). Accordingly, the corticofugal projection forms a tight loop of feedback between neurons of matched features in the subcortical auditory nuclei and the auditory cortex.

Although most cortico-collicular modulation experiments were conducted by cortical electrical stimulation in the bat, qualitatively the same effects on BF tuning were recorded in IC neurons regardless of the type of stimulation applied to the cortex, such as the repetitive tonal stimulation delivered to the animal, the focal electrical stimulation of auditory cortex, or the auditory fear conditioning . Auditory fear conditioning had nearly the same strength of effects as cortical electrical stimulation, but the tonal stimulation had less effects on the BF tuning (Gao and Suga 2000). Other aspects of auditory tuning in IC neurons, such as sensitivity to the duration of sound, the delay, and the minimum acoustic threshold, were similarly affected by cortico-collicular system. Therefore, the cortico-collicular modulation occurred on multiple representations of the sound feature and made widespread effects on the auditory perception and reorganization of variety of features in IC and cortex.

The time course of cortico-collicular effects on BF tuning was fast, and the effects were generally transient; but in the longest case the effects lasted several hours after the stimulation. The subcortical plastic changes were in general of shorter duration than the cortical change (Zhang and Suga 2000; Zhou and Jen 2005). Zhou and Jen (2000) demonstrated a brief cortical stimulation reduced or enhanced the sound-evoked responses of IC neuron just transiently (less than 5 min) or a short term (about 25 min). In extreme cases, the modulation lasted only for 10 s. In these IC neurons, the cortico-collicular projection appears to modulate and update the on-going processing of acoustical signals continuously.

Electrical stimulation of the cholinergic basal forebrain applied together with the auditory cortical electrical stimulation was effective to modulate BF tuning of cortical neurons (Bakin and Weinberger 1996; Bjordahl et al. 1998). The nucleus basalis is crucial for cortical plasticity through cholinergic disinhibition mechanisms (Casamenti et al. 1986; Rasmusson et al. 1992). The cortico-collicular modulation on the tuning of IC neurons depends on inhibition of response at certain frequency and facilitation of responses at other frequencies (Suga and Ma 2003). BF tuning in IC was further affected by local application of bicuculline (antagonist of $GABA_A$ receptors), namely, via disinhibition mechanism. Accordingly, the balance of excitation-inhibition may affect the tuning of IC neurons. Synaptic responses in IC neurons are well balanced between the excitation and inhibition in the mouse (Ono and Oliver 2014). It is of great interest to know the neural mechanism that operates to balance inputs between excitation and inhibition in individual IC neurons and particularly to know how the cortico-collicular descending activity affects the rebalancing of these synaptic inputs in IC neurons.

3.5 New Experimental Approaches

We have developed two approaches that could be useful in brain research of physiology and anatomy. One is to label neural projections using avian adeno-associated virus (A3V) that encodes fluorescence proteins. A3V transfects neurons anterogradely (Matsui et al. 2012). A3V-mCherry injected into the caudal telencephalon labeled corticofugal projection fibers a long distance to ipsilateral IC and has labeled further down to the brainstem nuclei. More interesting is the ascending projection that was labeled by A3V-mCherry injected into the IC. mCherry labeled neurons in the nOV (avian auditory thalamus), and furthermore the fibers and neurons in the telencephalon. Apparently, A3V-mCherry transfected neurons of nOV across the synapse terminal of IC projection fibers, and the transfected fibers in nOV projected to the telencephalon, and more neurons were transfected there.

Another unique approach is an invention of a photometric patch electrode (PME) recording system. PME uses a quartz glass patch electrode as a light guide to transmit laser light and excite fluorophores in neurons at the tip of electrode. The excited fluorescence is captured by the electrode and is transmitted back to a light detector, either a photomultiplier tube or a spectrometer. Since PME is an electrode, it can record simultaneously the neural electrical signals. Furthermore, PME can eject pharmacological agents locally from the tip of electrode.

3.5.1 *A3V-mCherry Visualized Ascending Auditory Projection Across Synapses*

We labeled ascending projection by injecting A3V-mCherry into the IC. After 7 days of injection, and immunostaining, mCherry-positive fiber projections were found lightly in the contralateral IC at the corresponding area to the ipsilateral injection site (Fig. 3.2a). Surprisingly, there were mCherry-positive neurons in nOV (nucleus ovoidalis, the auditory thalamus of avian species) and mCherry-positive fibers in the caudal telencephalon, the field L (Fig. 3.2b, c). Since A3V-mCherry was injected in the IC, presence of mCherry-positive cell bodies in nOV

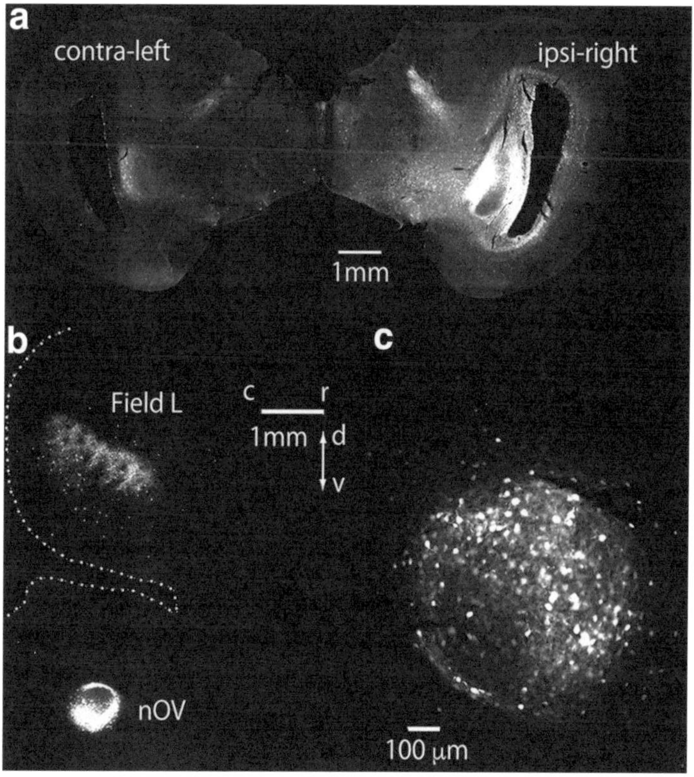

Fig. 3.2 Immunostaining of the ascending auditory projection from IC to telencephalon
(**a**) A3V-mCherry injection was made in the right IC where acoustic response was largest. Corresponding regions of contralateral IC were lightly mCherry positive
(**b**) Parasagittal section through the midbrain and the caudal telencephalon. Both field L and nOV are mCherry positive. Caudal circumference of the section is indicated by dotted line. In the midbrain, neurons in nOV distributed heterogeneous. In the caudal telencephalon, fiber projections in field L and several neurons were mCherry positive
(**c**) mCherry-positive neurons in nOV. Different preparation from B

and fiber projections in field L indicates that A3V transfected neurons across the synapse; across the afferent synapses between the fiber terminals from IC and neurons in nOV. Furthermore, the transfected neurons in nOV projected fibers long to field L. Figure 3.2b shows a parasagittal fluorescence image including regions of the midbrain and caudal telencephalon.

mCherry-positive neurons distributed irregularly in the nOV (Fig. 3.2b, c). Some areas in nOV were totally absent of mCherry-positive neurons, which may reflect sparse and possibly segregated projections from IC neurons to nOV. Heterogeneous organization of nOV is reported in the chicken (Wang et al. 2017).

3.5.2 A3V-mCherry Visualized Ipsilateral Descending Auditory Projection

Cortico-collicular modulation of auditory activity is important and seems underlying many auditory functions of higher orders. We were interested in that system and tried to label the projection using A3V-mCherry in the chicken. A3V-mCherry was injected in the caudal part of telencephalon where sound response was the largest. Figure 3.4b shows the injection site and the tract of injection pipette as a lining of mCherry-positive cells. Many mCherry-positive fibers radiated from the injection site and extended rostral in a dorsal route and also in a ventral route. Many mCherry-positive cells scattered around the injection site in the ipsilateral hemisphere. However, mCherry-positive neurons or fibers were not found in the contralateral hemisphere (Fig. 3.4a). Consistenly, major interconnecting structure was not reported between two cerebral hemishperes (Wang et al. 2010), although a presence of reentrant fiber projections was reported in field L and caudal mesopalium (CM) from the contralateral ventromedial nucleus of the intermediate archistriatum (Aivm, Fig. 3.3).

There were large clusters of mCherry-positive neurons in the caudal end of telencephalon (Fig. 3.4b). These regions could be the source of descending corticofugal projection. Aivm is reported as the source of auditory descending projection (see Wild et al. 1993; Wang et al. 2010). In the present experiments, A3V-mCherry was injected in a large quantity (~1 μL) in the telencephalon including the region of field L where acoustic response was robust. Thus, neurons responsible for the descending projection were likely transfected by A3V-mCherry leaked out from the injection site.

mCherry-positive fibers descended the ipsilateral side and surrounded the nOV, from the rostral to ventral perimeter of the nucleus (Fig. 3.3), and fine fiber branches appeared in the perinucleus region (Fig. 3.4c, d). Small mCherry-positive neurons scattered around the nOV. There were no projections to the contralateral nOV nor to the contralateral IC (Fig. 3.4c, e). The descending projection continued to the shell region of IC that surrounds the core region of IC (ICC) (Fig. 3.4e, f). Within ICC many mCherry-positive fibers with bouton-like structures branched and surrounded

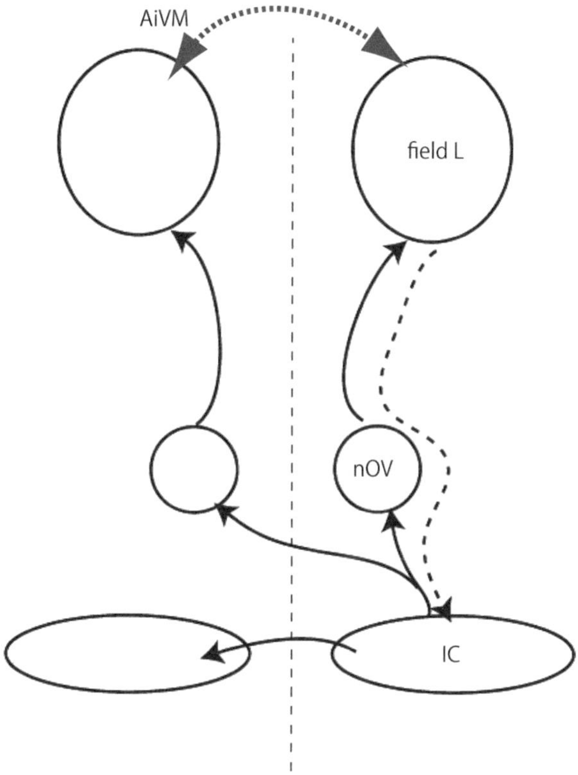

Fig. 3.3 Scheme of auditory feedback loops including field L, nOV, and IC
Projections from IC innervated contralateral IC, bilateral nOVs, and bilateral field L. Bilateral telencephalons are connected by reentrant fibers from AiVM (dotted line). Broken line indicates the cortico-collicular descending projection

hollow structures, perhaps neurons (Fig. 3.4g). Some mCherry-positive fibers extended projection to the tectum that surrounded the ventricle (Fig. 3.4e). Since neurons in ICC transmit ascending signals to bilateral nOVs, the presence of mCherry positive descending fiber projections in the shell of IC (Fig. 3.4f) apparently indicates that the core and shell of IC are the target of cortico-collicular descending projection and are the place of interaction between auditory ascending and descending signals.

Subcortical and cortical processing of auditory signals are modified and fine-tuned through the descending activity of corticofugal projection (Suga and Ma 2003). Because the descending projections are exclusively ipsilateral in avian, the modulation of neural processing is likely weighted toward the ipsilateral signal (Fig. 3.3, broken line); although the outcome of this biased descending innervation may be neutralized by repeated crossed innervations along the ascending pathway. The corticofugal projections are extending down to the cochlear organ. Thus, the auditory feedback system forms long loops which include all the subcortical auditory nuclei (mammals, Malmierca and Ryugo 2011). We need more detailed works to elucidate activity of all these corticofugal networks.

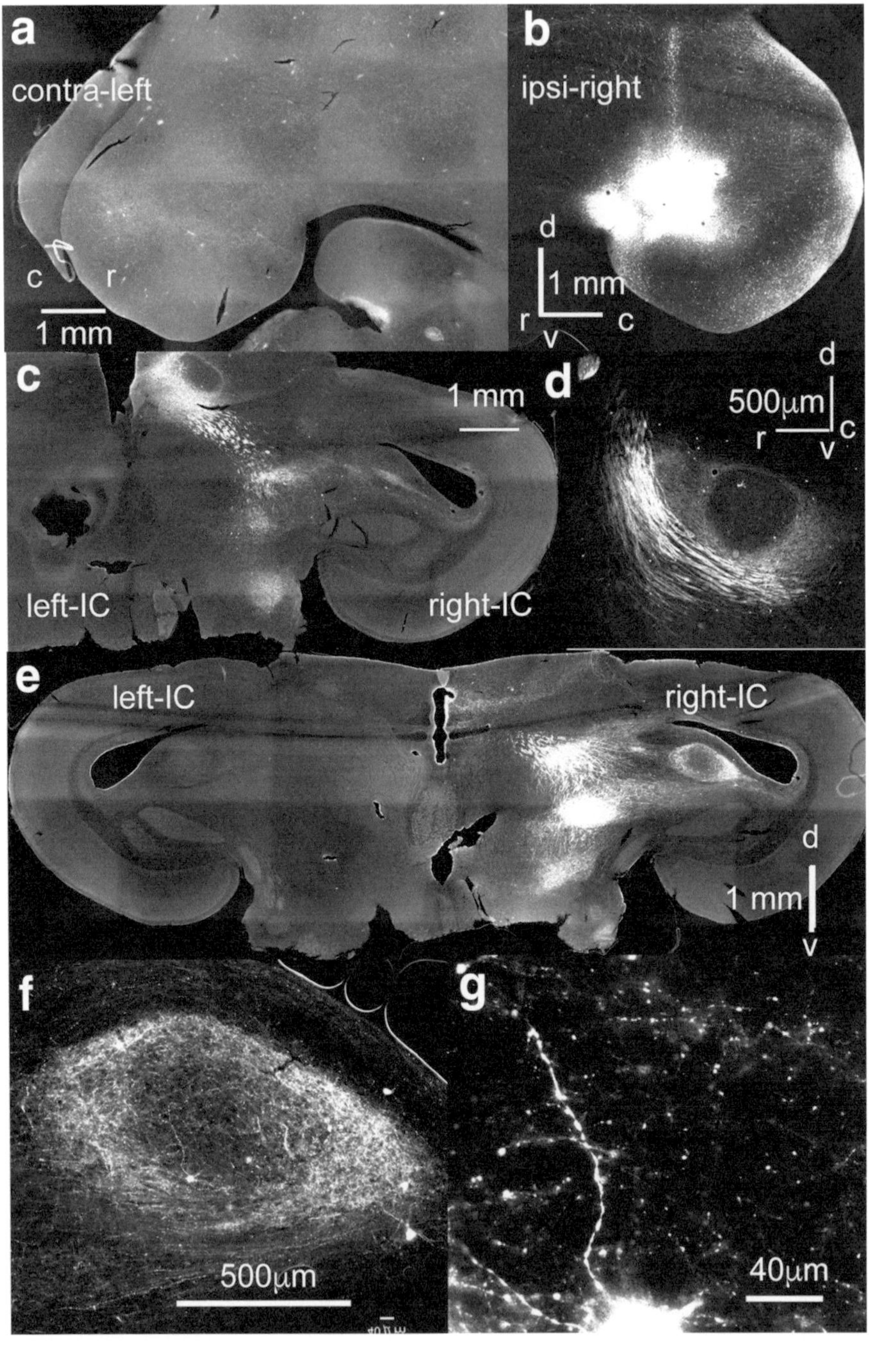

Fig. 3.4 Immunostaining of the descending projection from telencephalon to IC
(**a**) Parasagittal section of the left telencephalon, contralateral to the A3V-mCherry injection site

Systematic investigation of neural interactions could be possible by injections of A3V-fluorescence proteins of different immunogenic properties into different brain regions, such as the conversions of bilateral projections or the ascending and descending projections. One drawback to A3V is that the transfection capability is limited to the avian neurons only. We should challenge to find mammalian subtypes of AAV that could transfect neurons across the synapse. Moreover, combined application of such viral agents with commonly used anterograde or retrograde tracer dyes seems promising to visualize projections from different nuclei to single target neurons. Such labeling technology will be powerful to investigate neural networks on single cell basis.

3.5.3 PME Measures the Electrical and Optical Signals Simultaneously in the Brain In Vivo

Neural network is not simply a group of neurons interconnected by synapses, but it is the ensemble of neurons that provides a basis for synchronized and plastic activity. Activity of neural circuits is represented as electrical spikes of individual neurons and also mobilizes signaling molecules within neurons to maintain the cellular activity or to consolidate plastic changes. Spiking activity of neurons can be monitored by an electrode, while molecular activity can be monitored by optical signals. These two signals are generally measured by separate methods, such as the microelectrode for monitoring electrical activity and the confocal microscope for measuring distribution and dynamics of signaling molecules. Because two signals are generated by a common source but indicate different aspects of neural activity, simultaneous measurements of these two signals are ideal for the understanding of neural circuit activity.

We have challenged to develop a photometric patch electrode (PME) recording system which can measure the optical signal of neurons simultaneously with electrical activity (Hirai et al. 2015). PME is a quartz glass patch electrode and transmits laser light to excite fluorescence indicator molecules within cells and captures the

Fig. 3.4 (continued) (**b**) Parasagittal section of the right telencephalon where A3V-mCherry injection was made. The injection site was brightly immuno-positive. A track of injection electrode is visible as vertical cluster of mCherry-positive neurons

(**c**) Coronal section of IC. A3V-mCherry immunoreactivity was visible only in the ipsilateral IC. mCherry-positive fibers surrounded nOV passing through the medial peri-nucleus region

(**d**) Parasagittal section of right nOV, ipsilateral to the telencephalon injection site . Projection fibers passing the circumference of nOV from rostral to ventral perimeter. Small dots are mCherry positive

(**e**) mCherry-positive descending fiber projections were found only in the ipsilateral side of IC to the injection in telencephalon

(**f**) Ipsilateral IC region in an expanded scale. mCherry-positive fibers projected to the shell region of IC, and many fibers with bouton-like structures projected within the core

(**g**) Fiber projection from mCherry-positive neurons within ICC. Many mCherry-positive fibers with boutons surrounded IC neurons, the hollow structures without fluorescence

fluorescence signal at the tip of electrode. Moreover, PME can monitor signals not only from single cells as unit responses but also from cells ensemble as field responses and can apply chemicals locally from the tip (Hirai et al. 2015). Detailed methods of loading the indicator dyes or expression of genetic-encoded fluorescence molecules in the brain tissue and fabrication of PME will be described in the Appendix section of this chapter.

3.5.4 Basic Components of PME Recording System

PME recording system is composed of a part of laser transmission and a part of fluorescence detection (Fig. 3.5a). The laser-transmitting part consists of a laser-emitting assembly, one branch of a bifurcated optical fiber bundle to transmit the laser, and the PME holder assembly (Fig. 3.5b). Laser light is transmitted through the PME to excite fluorophores in neurons (Fig. 3.5c). Emitted fluorescence from neurons was captured at the tip of PME and was transmitted back to the other branch of bifurcated optic fiber bundle to fluorescence detectors. The fluorescence detecting part had two measuring options (Fig. 3.5a). One option uses a grating spectrometer (SPCT) that is coupled with a CCD camera to measure a full range spectrum of fluorescence signal (Fig. 3.5a-(i)). The other option uses a photomultiplier tube (PMT) and a lock-in amplifier to measure the fluorescence signal at fast speed (Fig. 3.5a-(ii)). The choice of detector depends on the type of experiment and the nature of fluorescence signal, whether the fluorescence signal has a fast response (choose the photomultiplier tube) or the signal includes a spectrum change (choose the spectrometer).

3.5.5 Fluorescence Excitation at the Tip of PME

Excitation of fluorescence was most effective at the tip of PME as is shown in Fig. 3.5c. The photograph is a brain slice prepared from rat cortex, in which some neurons expressed eGFP. There were several eGFP-positive neurons in this slice, and some neurons lightly emitted fluorescence because of the laser leak from the PME (see Fig. 3.11d). One neuron in the picture was touched to the tip of PME (indicated by a pair of broken lines) and emitted fluorescence intensely (Fig. 3.5c).

In a slice preparation, the laser leak from PME illuminated directly several eGFP-positive neurons. This happened under the situation where no light-scattering structures were present between the PME and eGFP-positive neurons. However, the laser light leakage from PME should be in a different condition in brain tissues in vivo, where the light scattering by cellular structures reduces the intensity of light for illumination of cells and for sensing the fluorescence. Accordingly, high-quality imaging by a multiphoton confocal scanning laser microscope is generally limited to the depth less than 1 mm from the brain surface (Helmchen and Denk 2005;

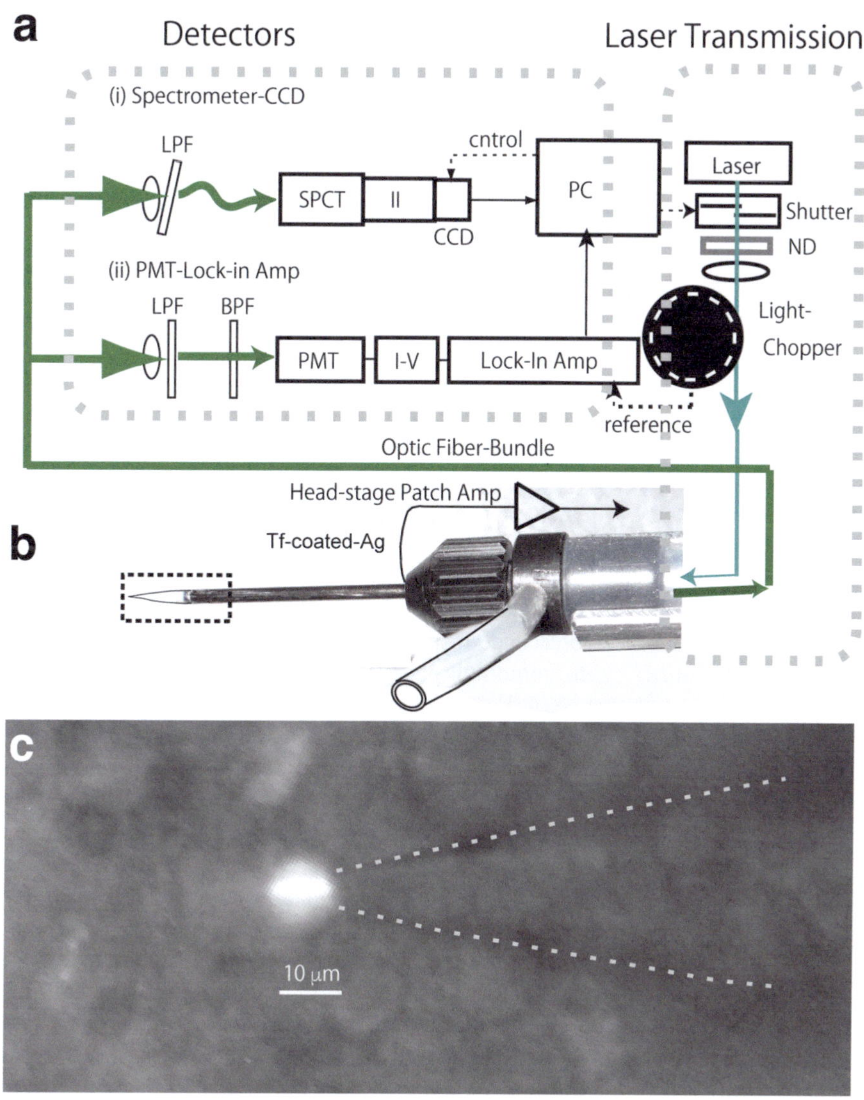

Fig. 3.5 Outline of photometric patch electrode (PME) recording system
(**a**) PME recording system is blocked in two parts. One part in the right is for laser transmission and the other in the left for light detections. Two modes of light detectors are illustrated. (i) CCD camera coupled with a spectrometer (SPCT) and (ii) photomultiplier tube (PMT) and a lock-in amplifier. In laser transmission part, laser light is transmitted to one branch of bifurcated optical fiber through a shutter, ND filter, focusing lens, and a light chopper that encodes reference signal to the lock-in amplifier. (**b**) PME and holder assembly. Ejection of liquid from the tip of PME is controlled by pressure within the PME through the tube. The tip region of PME is squared to be enlarged in C. (**c**) eGFP-transfected neurons in the brain slice of rat pup. Laser light was emitted from the tip of PME and excited fluorescence like a bar in the neuron contacted
Modified from Hirai et al. (2015)

Mittmann et al. 2011). In contrast the PME is a light guide that delivers the laser light at any depth of brain tissue and is a sensor that can detect fluorescence at the tip of electrode, which could overcome the depth limitation of fluorescence measurement.

3.5.6 Comparison of the Measurements of Calcium Signal Between Two Detectors in Slice Preparations: Spectrometer and Photomultiplier Tube

We compared the nature of Ca^{2+} signals between measurements that used the spectrometer (Fig. 3.5a(i)) and that used the photomultiplier tube (Fig. 3.5a(ii)) from slice preparations of the hippocampus of rat pup. Ca^{2+} signals were measured from OGB1 (Oregon green BAPTA1-AM)-loaded granule cells (loading procedure is detailed in Appendix of this chapter; Murayama ct al. 2007; Stosiek et al. 2003). PME touched the OGB1-loaded neuron in a slice preparation by positioning the PME under visual control (Fig. 3.6a inset). A series of electrical stimulation was applied to trigger antidromic spikes in a granule cell through a bipolar electrode (Fig. 3.6a, b). Calcium fluorescence response was measured by the spectrometer (Fig. 3.6c) and further by the photomultiplier tube (PMT) (Fig. 3.6d). Calcium signal was evaluated as a change of fluorescence level ΔF/F, where F is the basal fluorescence level and ΔF is the change of fluorescence. Calcium signals had a long decay time course extending for a few seconds, and the size was increased as a function of stimulus number in a train. The decay time course may reflect partly the accumulation of Ca ions within the cell by repetitive generation of antidromic spikes. Moreover, the decay time course of fluorescence signal may reflect the kinetics of Ca^{2+}-OGB1 binding and built-in buffering and exclusion mechanisms for calcium ions within granule cells (Helmchem et al. 1997; Chuhma et al. 2001).

Time resolution of calcium signal was better when a PMT was used as a detector (Fig. 3.6d). However, the response features such as duration of the signal and size dependence on stimulus number in a train were not different in these two measurements (Fig. 3.6c, d). Fluorescence signal was fast rising and noisier in the measurement by PMT, which reflected a fast sampling rate (5 ms). Signals detected by the spectrometer were sampled at slower rate because of the on-tip integration of light signals by CCD camera (0.3 s). Furthermore, the spectral signals were averaged for limited wavelength ranges (500–550 nm). All these procedures improved the signal to noise ratio of spectrometer signal (Fig. 3.6c) (Hirai et al. 2015).

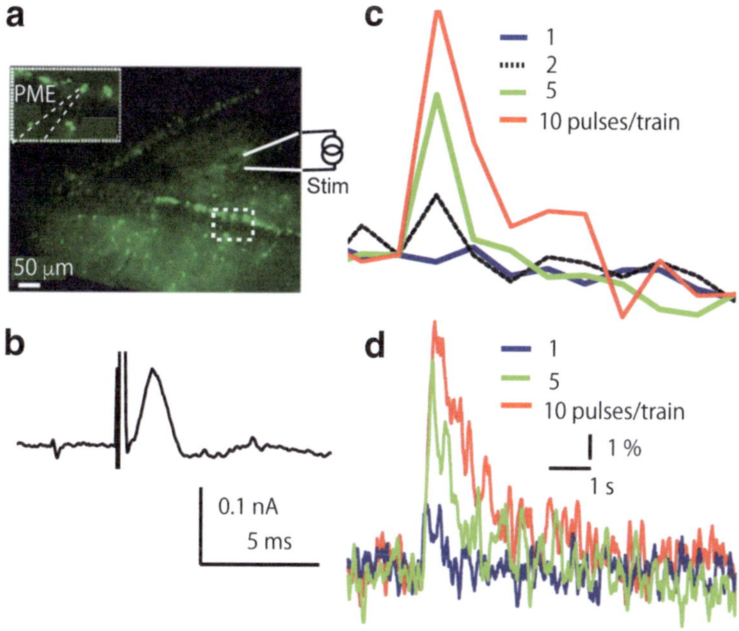

Fig. 3.6 Extracellular current signal and OGB1 calcium signal
(**a**) Granule cells were loaded with OGB1 in a slice of rat hippocampus. Antidromic electrical stimulation (Stim) was applied, and PME measurements were made. Inset shows expanded image of the squared region. PME is indicated by a pair of broken lines. (**b**) evoked response in the cell-attached PME recording. (**c**, **d**) calcium fluorescence signals evoked by electrical stimulus from 1 to 10 pulses per train. (**c**) spectrometer-CCD camera detector was used. (**d**) PMT lock-in amplifier detector was used. Note that time resolution is better in d but noise level is larger. Time resolution in c was 0.3 s and 5 ms in d. Calibrations in d are applied to both c and d

3.5.7 *In Vivo* Recordings from Neurons in Auditory Cortex Field L and in IC

Sound-evoked responses of neurons in field L and IC were measured in in vivo experiments from the chick. OGB1-AM was injected in these regions of brain while monitoring acoustic responses. After 1 h of OGB1-AM injection, a PME was inserted to the injection site. OGB1-AM is a membrane-permeable compound and is cleaved by esterase activity within a cell to produce the Ca^{2+}-sensitive OGB1, which binds with Ca ions and emits fluorescence by excitation. Figure 3.7 illustrates the case of OGB1 injection to field L. The depth of fluorescence recording was about 3 mm from the brain surface. Injection and recording from the IC followed the same procedure except for the depth, which was about 7 mm from the brain surface. PME was inserted to the OGB1-AM injected site. When approaching to the injected site, local field electrical responses to stimulus sound were monitored. The recording site was also confirmed by the increase of OGB1 fluorescence signal

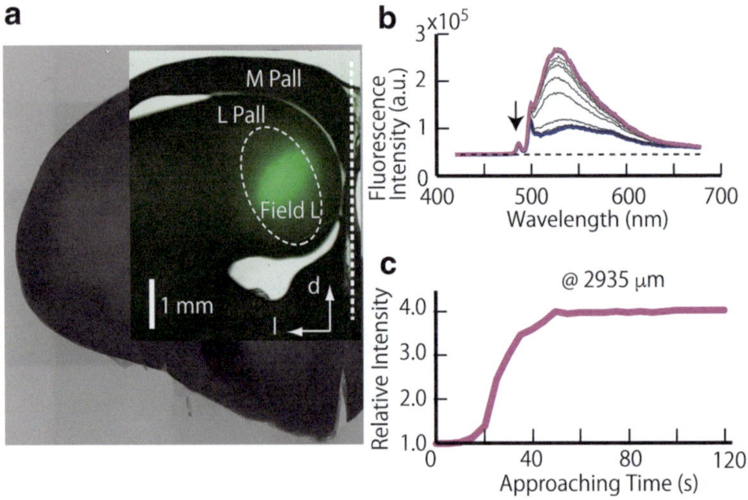

Fig. 3.7 OGB1-labeled field L
(**a**) Slice preparation of caudal telencephalon to show OGB1 injected site near the field L. Field L is located at the depth of 3–4 mm from the brain surface. (**b**, **c**) approach of PME to the OGB1 injected site increased the fluorescence signal, as the spectrum (**b**) and the peak level of the spectrum (**c**). Blue line indicates the basal spectrum measured at the surface of the telencephalon, and the magenta indicates the maximum spectrum intensity. The fluorescence signal was maximum at the depth of 2935 μm
Modified from Hirai et al. (2015)

(Fig. 3.7b, c). When a spectrometer was used as a detector, fluorescence intensity of the wavelength ranging from 500 to 600 nm increased up to 300–600% of the starting fluorescence level that was the measure on the brain surface (Hirai et al. 2015). With a PMT also, the increase of basal fluorescence level was used to confirm the injection site.

Electrical activity of field L is unique to show spontaneous low-frequency oscillation (Fig. 3.8). The oscillation appeared like the slow field potential oscillation in the mammalian neocortical neurons (Steriade et al. 1993) and took upstates and downstates. From the upstates spikes were generated (Fig. 3.8a), and spikes accompanied with calcium signals (Fig. 3.8b). These calcium signals were measured by PMT. Origin of these spontaneous oscillation is not yet identified.

Sound-evoked unit activity was robust in IC neurons. However, the fluorescence signals were small so that spectrometer was used to detect these signals. The spectrometer is more sensitive and has a better signal to noise ratio than PMT, through the signal amplification by the image intensifier and integration of spectrum signals by the CCD camera (Fig. 3.5a); however, the time resolution of fluorescence signal was not great (0.3 s). Unit spike activity was sustained during a sound stimulus (Fig. 3.9, green bars, left), which accompanied with fluorescence response. The fluorescence response enveloped a series of unit activity (black solid line, Fig. 3.9

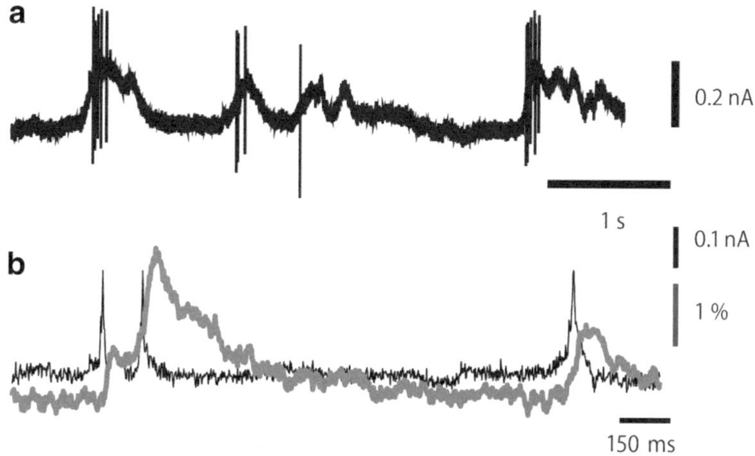

Fig. 3.8 Spontaneous local field current activity in field L
(**a**) Upstate and downstate. Spikes were generated from the upstate. (**b**) local field electrical activity (solid black line) and calcium response (gray line) from OGB1-labeled field L

right), and the size was nearly constant regardless of the sound duration and number of spikes. This likely indicated that calcium ions were not accumulated intracellularly in IC neurons despite of the increase of spike numbers. This is in contrast to the increase of fluorescence signal in granule cells in hippocampus with the spike number in a train (Fig. 3.6). Thus, the observations may indicate a presence of robust calcium buffering activity and exclusion mechanisms within the IC neuron.

3.5.8 Correlation Between Electrical Activity and Calcium Signal

As expected, local field current and fluorescence response to sound stimulus were highly correlated both in time course and amplitudes (Fig. 3.10a1, a2), when signals were sampled by a photomultiplier tube at fast rate (PMT mode). Here, acoustic responses in field L were enhanced by gabazine (0.1–5 mM) that was applied locally through the recording PME. Gabazine was pressure applied from the tip of PME and blocked the inhibitory synapse and enhanced the sound-evoked responses. Calcium response followed local field current with a delay of 50–80 ms (Fig. 3.10a). This relatively long delay of calcium response to track electrical signal may reflect that both signals were originated in the ensemble activity of local neurons. Figure 3.10b illustrates scatter plots between peaks of local field current and fluorescence signal measured simultaneously. There are two clusters of signals: small local field currents accompanied with small fluorescence responses and larger currents with larger fluorescence responses. The scatter plots for larger signals had a

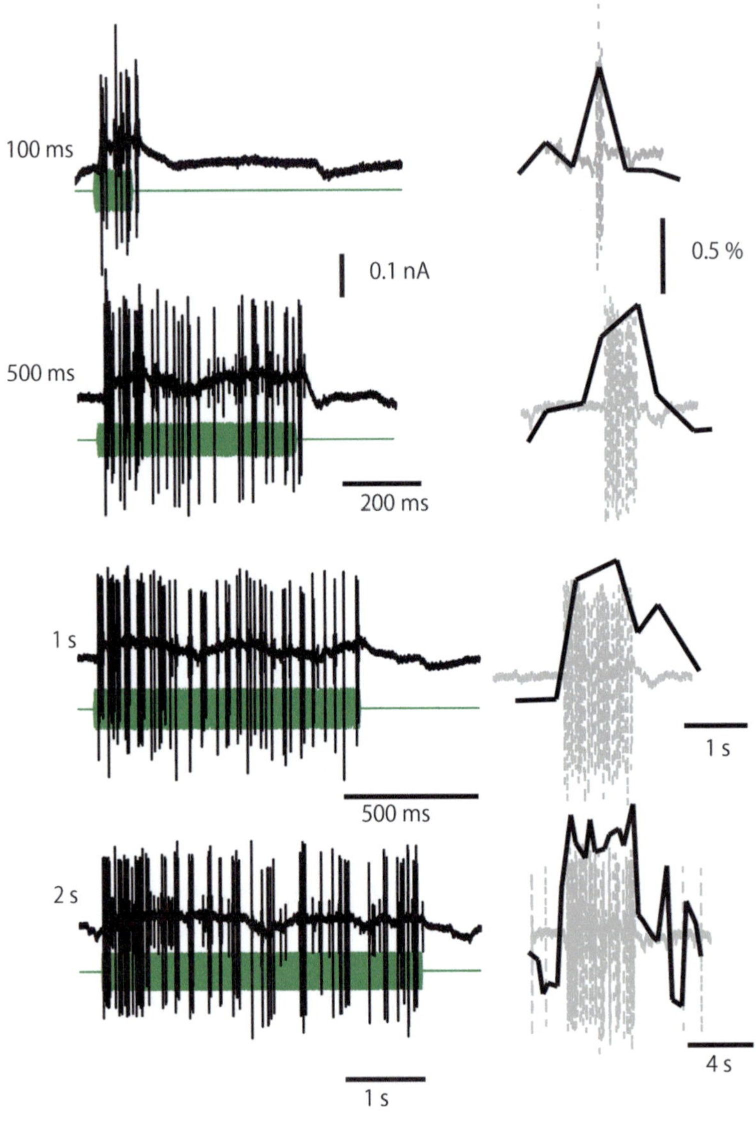

Fig. 3.9 Unit activity evoked by sound stimulus in IC
IC was loaded with OGB1. In the left, sound stimuli were illustrated as bars (green) and the sound-evoked unit activity as solid black line. Unit activity and fluorescence responses were measured using spectrometer-CCD camera system. In the right, unit activity was compressed along the time axis (gray line), and calcium response (solid black line) was overlaid on spike responses
Modified from Hirai et al. (2015)

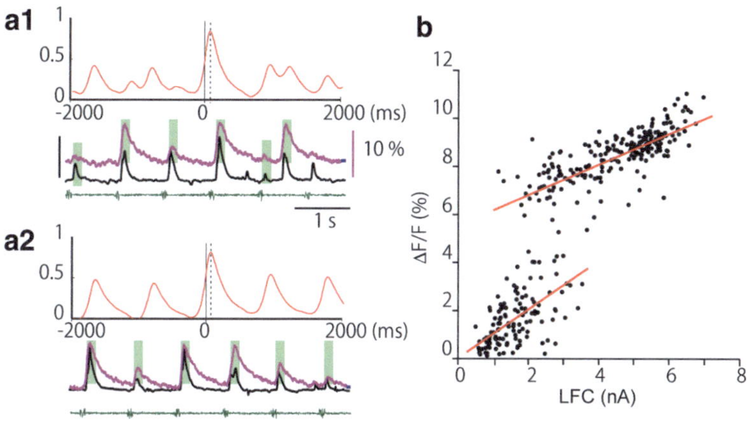

Fig. 3.10 Cross-correlation between the local field current and fluorescence calcium signals in field L
(**a1**, **a2**) two examples illustrate the cross-correlation (top, orange line) between the calcium signals (upper middle, magenta line) and current signals (lower middle, black line). Local field current and fluorescence signals were evoked by sound stimulus (indicated at the bottom). Light green vertical bars indicate simultaneous occurrence of local field calcium and current signals. Signal amplitudes were enhanced by local application of gabazine (0.5 mM). Calcium signals delayed to current signals by 70–80 ms. (**b**) scatter plots of peak amplitudes of fluorescence signal and current signal measured within regions labeled by green bars in **a**. There are two clusters. Small current signals ($r = 0.58$) were likely from subthreshold responses, and large current signals ($r = 0.76$) were from suprathreshold responses
Modified from Hirai et al. (2015)

shallower slope than those of smaller signals. The size of large fluorescence response was nearly saturated at ~10% (ΔF/F). Two response groups may correspond to subthreshold and suprathreshold neural activity, respectively (Fig. 3.10b).

3.5.9 Improving the Signal to Noise Ratio of Fluorescence Measurement

There is a leak of laser light from the side wall of PME (Fig. 3.11d). This light leak occurs mostly from the glass wall corresponding to the sharp taper of electrode, where the transmitted light inside PME strikes the glass wall with a large angle and leaks out from the surface. The laser leak is visible as light shower when the PME is dipped in a solution dissolving a fluorescence dye (Alexa Fluor 488) (Fig. 3.11d). The laser light leakage might generate noise in the fluorescence signal and will reduce the signal to noise ratio. Nevertheless, this disadvantage from the light leakage could not be so serious in the deep brain tissue in vivo, because of the scattering of light in the live tissues. As a matter of fact, the PME has succeeded in measuring Ca^{2+} fluorescence responses that were highly correlated with the electrical activity

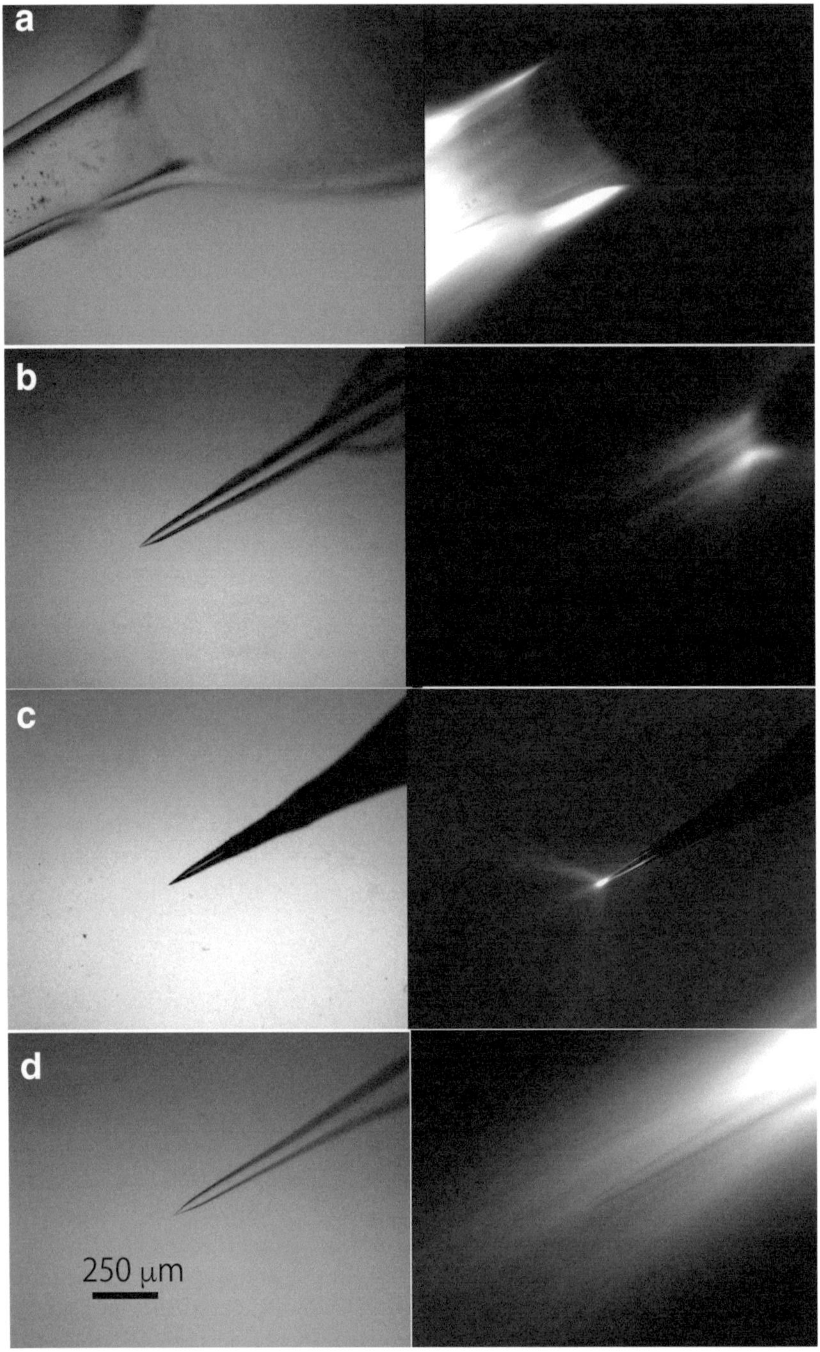

Fig. 3.11 Sylgard coating of PME and laser emission
(**a**–**d**), bright field images in the left and laser emission images in fluorescent solution (Alexa Fluor 488) in the right. Sylgard coating was made over the tapered region of the electrode (**a**) and close

in field L in vivo (Fig. 3.10). However, it should be noted that we have enhanced the signal to noise ratio of sound-evoked responses in these experiments by locally applying gabazine from the tip of PME (Fig. 3.10). Gabazine blocks $GABA_A$ receptors and enhances both electrical and optical signals of local field responses. Because the block of inhibitory synapses likely modifies the normal network activity, it is ideal to enhance signal to noise ratio without using these pharmacological agents.

There are alternatives to enhance fluorescence responses and to improve the signal to noise ratio without suppressing inhibitory synapses. (1) An increase of the power of excitation laser could be one alternative. One possible drawback of a large power laser could be an increase of optical noise due to the intensified laser light leakage and acceleration of bleaching of indicator compounds. (2) Another possibility seems to excite a larger mass of neurons by using a broad beam of laser light. Among several methods to achieve a broad laser beam, we adopted coating the PME with a resin that has optical refractive index smaller than that of quartz PME. This is also a technology to be used in telecommunication to prevent leakage of light from the optic fiber by coating the core with a clad. We coated the elongated part of PME close to the tip with Sylgard (Sylgard 184, Dow Corning). Sylgard is a familiar resin to coat the surface of patch electrode to reduce capacitive coupling of electrode against the ground in bathing medium.

Sylgard coating reduced leakage of laser light from the wall of PME, but still strong light was emitted from the coated end of Sylgard (Fig. 3.11a, b). Sylgard has smaller refractive index (1.4–1.43) than a quartz glass (1.5–1.55). Accordingly, Sylgard coating kept the laser light within the quartz glass wall. The laser light is still emitted from the end of Sylgard coat of PME because the light beam crosses the surface of coated Sylgard almost at straight angle (Fig. 3.11a, b, right). Thus, a population of neurons near the tip of PME could be illuminated by the laser. Graded index optic glass is another example that has analogous mechanism to keep the light transmission within the glass by arranging the refractive index gradually lower toward the perimeter of the glass. When Sylgard was mixed with small particles like soot, the laser emission occurred only at the very tip of PME (Fig. 3.11c). Soot in the Sylgard absorbs the light and blocks the laser leak and blocks even the transmission of light through the resin. Even with Sylgard coating, the PME measures electrical signal in high resolution, and can eject chemicals locally under a pressure control. Thickness of Sylgard coating is flexible. With a thick coating, we can illuminate relatively a large area and can excite cells ensemble near the tip of PME, while with soot in Sylgard, we can reduce the light emission from the tip to the minimum that seems appropriate to excite and record a single cell activity.

Fig. 3.11 (continued) to the tip of electrode (**b**). Sylgard coating prevented emission of laser light from the wall of PME, but strong light was emitted from the end of Sylgard coat. (**c**) When Sylgard was mixed with soot and coated the PME close to the tip, laser leakage was totally protected, and light emission occurred only at the tip of PME. (**d**) laser light emission from PME that was not Sylgard coated Modified from Hirai et al. (2015)

3.5.10 *Expression of Genetic-Encoded Calcium Indicators (GECI)*

PME measurements in combination with genetic-encoded calcium indicator (GECI) proteins are ideal, because we can express the protein in targeted neurons or brain regions. This technique allows us recording from special source of neurons or a brain region and also at controlled timing when combined with the conditional expression technique. However, a sensitive indicator protein such as GCaMP6 has generally a low-level basal fluorescence and makes it difficult to search the expression site in deep brain regions (Akerboom et al. 2012). This drawback can be slightly overcome by a long incubation time after the viral injection, which is to aim an increase of number of GECI expressing cells. With GCaMP6 encoded in AAV we maintained the animal at least a few weeks or longer period of time after the injection to be used in experiments. During penetration of PME, a small increase of basal fluorescence signal (~10%) was detected, which indicated the expression site and we could record substantial Ca^{2+} fluorescence responses there. This low-level signal increase at the injection site may reflect the low-level basal fluorescence of GCaMP6 without bound Ca ions. Presently, we used only the calcium indicator compounds with PME to measure neural activity. Indicators could be extended to report other interesting signaling molecules, such as GPCRs, G-protein-coupled receptors.

3.6 Recapitulation of this Chapter

Cortical neural activity is plastic normally in juvenile animals before closing the critical period. Moreover, the auditory cortex becomes plastic even in adult animals in a specific environmental condition. Chronic exposure to mild noise sound reopens the critical period in adult animals (in the rat). Even in natural environments, auditory cortex of some songbirds becomes plastic every year seasonally. Details of song learning network activity were investigated in variety of song birds. It is known that sensitivity of sound source localization of the barn owl is controlled by visual signal. Fiber projections within IC are reoriented by redirected visual signals in the owl that wears spectacles. Moreover, the frequency tuning property of IC neurons in adult bats is modulated by corticofugal projection, and the activity of neurons of matched feature (i.e., frequency sensitivity) in IC and in the auditory cortex are fine-tuned. The importance of descending innervations in auditory signal processing is started to be addressed experimentally. The interaction between the ascending and descending signals seems crucial in the neural network to fine-tune the processing of auditory information; special example is the sensation to the pitch. Absolute pitch sensation seems innate to some people, but most people obtain it through training or do not have it at all. Presently we do not know any of the corresponding neural mechanisms of the pitch sensation. Labeling neural projections may facilitate the challenge to the understanding of neural network activity. Also the new electrode recording technique, the PME could be used to analyze the complex neural activity by recording simultaneously the electrical signal and the optical signal at any depth of the brain in vivo. The electrical signal, the spike, represents the output of neural activity and the optical signal reports underlying metabolic processes. Metabolic processes in neurons maintain ongoing neural activity and reform neural circuits through mobilizing variety of enzymatic activity including gene expressions. PME is a unique device, and its application in brain research has just recently started.

Appendix

Labeling Brain Regions In Vivo by OGB1 or by Avian Adeno-Associated Virus Encoding mCherry

Oregon Green BAPTA1-AM (1 mM in ACSF, OGB1-AM, Molecular Probes, Eugene, OR) was injected in brain regions about 1 h ahead of PME recording. OGB1-AM was prepared by dissolving in 20% Pluronic F127 in DMSO and diluted in ACSF (concentrations in mM: 125 NaCl, 2.5 KCl, 26 NaHCO3, 1.25 NaH_2PO_4, 2 $CaCl_2$, 1 $MgCl_2$, and 17 glucose, pH 7.4) (Grienberger and Konnerth 2012; Murayama et al. 2007; Stosiek et al. 2003). The glass electrode for injection was pulled from a hard glass capillary (1.5 mm OD, 1.0 mm ID) and had a tungsten wire (15 μm, Teflon coated) inside to monitor the sound-evoked electrical field response. The dye injection was made at the point where acoustic response was the largest. The tip of glass electrode was cut to make 20–30 μm diameter. The tungsten wire was also cut at the level of capillary tip and was made sensitive to neural electrical activity just at the exposed tip. The tungsten wire was fastened to the electrode at the proximal end by bending and thrusting through a silicone tube. Since the wire was Teflon coated for insulation, the coat must be removed either by cigarette lighter flame or by scraping at the point where the wire contacted with the electric lead wire to an amplifier. The electrode was filled with a small amount of dye solution at the tip (1–2 μl).

A minimum level of positive pressure was applied manually to eject the dye solution out from the tip of glass electrode, by using a subcutaneous injection syringe (1 mL). We maintained the positive pressure inside the capillary by using a three-way valve for 30–60 s. At the moment of dye ejection, the acoustic field response was reduced transiently, but the response recovered quickly after the ejection (within 30 s) in most cases. Therefore, we believe that the ejected volume was small. The pressure was released, and the electrode was maintained at that position further for approximately 3 min and was then withdrawn slowly.

Adeno-associated virus (AAV) is a nonpathogenic single-stranded DNA virus that can be replicated only in the presence of a helper virus such as herpes virus or adeno virus (Buller et al. 1981). The recombinant AAV vector permits nontoxic transduction of postmitotic cells and long-term gene expression in neurons but not in non-neuronal cells, which enables us to deliver tools for visualizing and manipulating neural activity. mCherry encoded in the avian strain of adeno-associated virus (A3V, 10^{13} particles/ml) was a gift from Drs. R Matsui and D Watanabe, Kyoto University. Details of A3V vector are described in their document (Matsui et al. 2012).

A3V-encoded mCherry was injected in the chick brain aged 2–3 days after hatch (P2–3) by the same protocol used for OGB1-AM injection, at the depth where the acoustic response was the largest. After the injection, a piece of craniotomy bone was returned to cover the brain surface, and cranial skin was sutured. After 7–10 days from viral transfection, the chick was deep anesthetized, whole brain was fixed in a fixation medium, and field L and IC were sliced at 50 μm thickness in agarose gel (4%). Slices were immuno-stained and mounted on slide glasses with vector shield. A series of fluorescent images were obtained by a fluorescence microscope using a x4 objective (BZ-9000, Keyence Japan). Adjacent frames were automatically merged to make a wide view of field L and IC images (Fig. 3.2, 3.4).

Fabrication of PME

PME was pulled from a quartz glass capillary. The exterior of PME was nickel (Ni) coated except at the center 10 mm and both ends of the capillary (Appendix-Fig. 3.12a) (Hirai et al. 2015). Both ends of the quartz capillary were polished by an optic fiber polishing machine (SpecPro Polisher; Krell Technologies, Westford, MA) to increase transmission of light at the interface with optic fiber bundle (Fig. 3.12b). The PME was pulled by a laser puller (P-2000/G; Sutter Instrument, Novato, CA) and had a tip diameter of 2–3 μm and a resistance of 1–3 Mohm when filled with ACSF. The tapered tip region of PME was occasionally coated with Sylgard (Sylgard 184, Dow Corning) to prevent leakage of laser light (Fig. 3.11).

PME Holder Assembly

Proximal polished end of PME apposed closely to the distal end of optic fiber bundle within the electrode holder assembly (Fig. 3.13). Electrical signal from the PME was led by a Teflon-coated silver (Tf-Ag) wire directly to the head stage of a patch clamp amplifier. The wire came out of the holder through the same hole with PME. Teflon coating of Ag wire was necessary to protect the wire from contacting the nickel-coated surface of the electrode. Teflon coat was removed by cigarette lighter flaming at both ends of the wire.

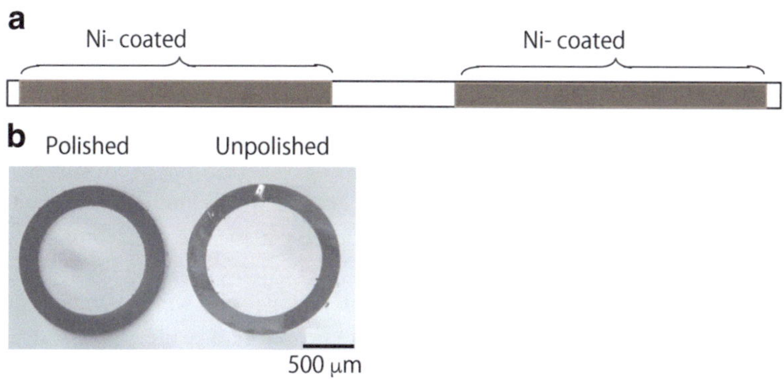

Fig. 3.12 Quartz capillary of PME
(**a**) Ni coating was made to the capillary. (**b**) both ends of the capillary were polished by a glass fiber polishing machine (left polished and right before polishing)
Modified from Hirai et al. (2015)

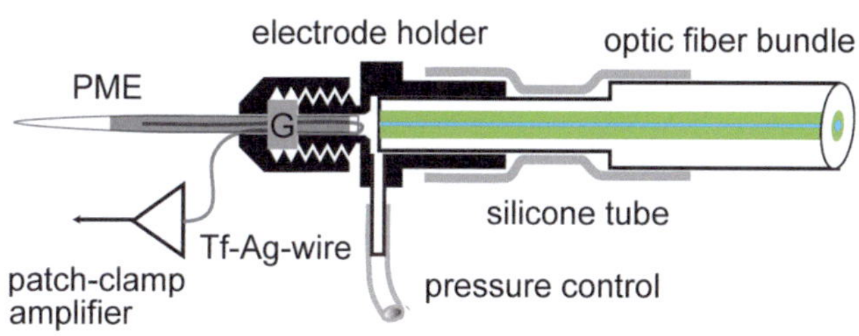

Fig. 3.13 Electrode holder assembly
Within the holder, optic fiber apposed PME directly. Pressure within the PME was controlled and regulated the fluid ejection from the tip. Electrical signal was led by Teflon (Tf)-coated Ag wire to a patch clamp amplifier. G, gasket
Modified from Hirai et al. (2015)

Bifurcated Optic Fiber Bundle

Bifurcated optic fiber bundle was used to separate the fluorescence signal from excitation laser light (Fig. 3.5). Ten percent of the fiber bundle transmitted the excitation laser, and the rest 90% captured the fluorescence and transmitted it to the light detectors.

Grating Spectrometer

Two types of light detectors were switched manually (Fig. 3.5a). One detector was a grating spectrometer (SPCT), and the other was a photomultiplier tube (PMT). Spectrum of fluorescence signal was obtained by the grating spectrometer, image intensifier, and CCD camera in series (Fig. 3.5a-(i)). Image intensifier was gated by a trigger that was issued from the PC (Fig. 3.5a). By using the spectrometer, optical signals from the detector probe were dispersed and expanded along the wavelengths (450–750 nm). Individual optic fibers of the detector probe were vertically aligned at the input to the spectrometer and fluorescence signals were dispersed against the wavelength. Thus aligned spectrum image was projected to II (image intensifier), intensified and was captured as an image by a CCD camera. The spectrum signal was on chip integrated by the CCD camera, in most experiments for 0.3 s (with 2 × 2 binning). A single spectrum profile was calculated from these spectrum lines. The power of spectrum was calculated by averaging over limited wavelength range (530–570 nm for OGB1). These procedures increased the signal to noise ratio of fluorescence measurement, but the time resolution was limited by the time used for on-chip integration by the CCD camera.

Photomultiplier tube (PMT) and a lock-in amplifier measured signals at faster speed. Photo-current output was current-voltage converted (LI-76, NF corporation, Yokohama, Japan), low-pass filtered through a lock-in amplifier (LI-5640, NF corporation), and analogue to digital converted (NI-DAQm; National Instruments; Fig. 3.5a-(ii)). By using a light chopper (50% duty cycle, 5584A, NF corporation), reference signals (~2 kHz) were provided to the lock-in amplifier (Fig. 3.5a). Optical signal was band-pass filtered before the PMT (BPF, 525–557 nm, Semrock FF01–542).

References

Aamodt SM, Kozlowski MR, Nordeen EJ, Nordeen KW (1992) Distribution and developmental change in [3H]MK-801 binding within zebra finch song nuclei. J Neurobiol 23(8):997–1005

Akerboom J, Chen TW, Wardill TJ, Tian L, Marvin JS, Mutlu S, Calderón NC, Esposti F, Borghuis BG, Sun XR, Gordus A, Orger MB, Portugues R, Engert F, Macklin JJ, Filosa A, Aggarwal A, Kerr RA, Takagi R, Kracun S, Shigetomi E, Khakh BS, Baier H, Lagnado L, Wang SS, Bargmann CI, Kimmel BE, Jayaraman V, Svoboda K, Kim DS, Schreiter ER, Looger LL (2012) Optimization of a GCaMP calcium indicator for neural activity imaging. J Neurosci 32(40):13819–13840

Akutagawa E, Konishi M (2010) New brain pathways found in the vocal control system of a songbird. J Comp Neurol 518(15):3086–3100

Arnold AP, Bottjer SW, Brenowitz EA, Nordeen EJ, Nordeen KW (1986) Sexual dimorphisms in the neural vocal control system in song birds: ontogeny and phylogeny. Brain Behav Evol 28(1–3):22–31

Aronov D, Andalman AS, Fee MS (2008) A specialized forebrain circuit for vocal babbling in the juvenile songbird. Science 320(5876):630–634

Bakin JS, Weinberger NM (1996) Induction of a physiological memory in the cerebral cortex by stimulation of the nucleus basalis. Proc Natl Acad Sci U S A 93(20):11219–11224
Basham ME, Nordeen EJ, Nordeen KW (1996) Blockade of NMDA receptors in the anterior forebrain impairs sensory acquisition in the zebra finch (Poephila guttata). Neurobiol Learn Mem 66(3):295–304
Bauer EE, Coleman MJ, Roberts TF, Roy A, Prather JF, Mooney R (2008) A synaptic basis for auditory-vocal integration in the songbird. J Neurosci 28(6):1509–1522
Berardi N, Pizzorusso T, Maffei L (2000) Critical periods during sensory development. Curr Opin Neurobiol 10(1):138–145
Bjordahl TS, Dimyan MA, Weinberger NM (1998) Induction of long-term receptive field plasticity in the auditory cortex of the waking guinea pig by stimulation of the nucleus basalis. Behav Neurosci 112(3):467–479
Boettiger CA, Doupe AJ (2001) Developmentally restricted synaptic plasticity in a songbird nucleus required for song learning. Neuron 31(5):809–818
Bottjer SW, Johnson F (1997) Circuits, hormones, and learning: vocal behavior in songbirds. J Neurobiol 33(5):602–618
Bottjer SW, Miesner EA, Arnold AP (1984) Forebrain lesions disrupt development but not maintenance of song in passerine birds. Science 224(4651):901–903
Brainard MS, Doupe AJ (2002) What songbirds teach us about learning. Nature 417(6886):351–358
Brainard MS, Knudsen EI (1998) Sensitive periods for visual calibration of the auditory space map in the barn owl optic tectum. J Neurosci 18(10):3929–3942
Buller RM, Janik JE, Sebring ED, Rose JA (1981) Herpes simplex virus types 1 and 2 completely help adenovirus-associated virus replication. J Virol 40(1):241–247
Bullock TH, Hagiwara S (1957) Intracellular recording from the giant synapse of the squid. J Gen Physiol 40(4):565–577
Buonomano DV, Merzenich MM (1998) Cortical plasticity: from synapses to maps. Annu Rev Neurosci 21:149–186
Casamenti F, Deffenu G, Abbamondi AL, Pepeu G (1986) Changes in cortical acetylcholine output induced by modulation of the nucleus basalis. Brain Res Bull 6(5):689–695
Chuhma N, Koyano K, Ohmori H (2001) Synchronisation of neurotransmitter release during postnatal development in a calyceal presynaptic terminal of rat. J Physiol 530(Pt 1):93–104
DeVoogd T, Nottebohm F (1981) Gonadal hormones induce dendritic growth in the adult avian brain. Science 214(4517):202–204
Doupe AJ, Solis MM, Kimpo R, Boettiger CA (2004) Cellular, circuit, and synaptic mechanisms in song learning. Ann N Y Acad Sci 1016:495–523
Elliott TM, Theunissen FE (2011) Chapter 20: The avian auditory pallium. In: Winer JA, Schreier CE (eds) The auditory cortex. Springer, New York, pp 429–442
Fagiolini M, Hensch TK (2000) Inhibitory threshold for critical-period activation in primary visual cortex. Nature 404(6774):183–186
Fagiolini M, Fritschy JM, Löw K, Möhler H, Rudolph U, Hensch TK (2004) Specific GABAA circuits for visual cortical plasticity. Science 303(5664):1681–1683
Feldman DE, Knudsen EI (1994) NMDA and non-NMDA glutamate receptors in auditory transmission in the barn owl inferior colliculus. J Neurosci 14(10):5939–5958
Froemke RC, Merzenich MM, Schreiner CE (2007) A synaptic memory trace for cortical receptive field plasticity. Nature 450(7168):425–429
Furukawa T (1981) Effects of efferent stimulation on the saccule of goldfish. J Physiol 315:203–215
Galambos R (1956) Suppression of auditory nerve activity by stimulation of efferent fibers to cochlea. J Neurophysiol 19(5):424–437
Gao E, Suga N (2000) Experience-dependent plasticity in the auditory cortex and the inferior colliculus of bats: role of the corticofugal system. Proc Natl Acad Sci U S A 97(14):8081–8086
Grienberger C, Konnerth A (2012) Imaging calcium in neurons. Neuron 73:862–885
Helmchem F, Borst JG, Sakmann B (1997) Calcium dynamics associated with a single action potential in a CNS presynaptic terminal. Biophys J 72(3):1458–1471

Helmchen F, Denk W (2005) Deep tissue two-photon microscopy. Nat Methods 2(12):932–940

Hirai Y, Nishino E, Ohmori H (2015) Simultaneous recording of fluorescence and electrical signals by photometric patch electrode in deep brain regions in vivo. J Neurophysiol 113(10):3930–3942

Hodgkin AL, Huxley AF, Katz B (1952) Measurement of current-voltage relations in the membrane of the giant axon of Loligo. J Physiol 116(4):424–448

Huang ZJ, Kirkwood A, Pizzorusso T, Porciatti V, Morales B, Bear MF, Maffei L, Tonegawa S (1999) BDNF regulates the maturation of inhibition and the critical period of plasticity in mouse visual cortex. Cell 98(6):739–755

Hyde PS, Knudsen EI (2002) The optic tectum controls visually guided adaptive plasticity in the owl's auditory space map. Nature 415(6867):73–76

Iyengar S, Viswanathan SS, Bottjer SW (1999) Development of topography within song control circuitry of zebra finches during the sensitive period for song learning. J Neurosci 19(14):6037–6057

Jen PH, Zhou X (2003) Corticofugal modulation of amplitude domain processing in the midbrain of the big brown bat, Eptesicus fuscus. Hear Res 184(1–2):91–106

Keller GB, Hahnloser RH (2009) Neural processing of auditory feedback during vocal practice in a songbird. Nature 457(7226):187–190

Knudsen EI (2002) Instructed learning in the auditory localization pathway of the barn owl. Nature 417(6886):322–328

Knudsen EI, Konishi M (1978) Space and frequency are represented separately in auditory midbrain of the owl. J Neurophysiol 41(4):870–884

Knudsen EI, Konishi M (1979) Mechanisms of sound localization in the barn owl (Tyto alba). J Comp Physiol 133:13–21

Knudsen EI, Knudsen PF, Esterly SD (1984) A critical period for the recovery of sound localization accuracy following monaural occlusion in the barn owl. J Neurosci 4(4):1012–1020

Leonardo A, Konishi M (1999) Decrystallization of adult birdsong by perturbation of auditory feedback. Nature 399(6735):466–470

Luo M, Perkel DJ (1999) A GABAergic, strongly inhibitory projection to a thalamic nucleus in the zebra finch song system. J Neurosci 19(15):6700–6711

Luo M, Ding L, Perkel DJ (2001) An avian basal ganglia pathway essential for vocal learning forms a closed topographic loop. J Neurosci 21(17):6836–6845

Ma X, Suga N (2001) Plasticity of bat's central auditory system evoked by focal electric stimulation of auditory and/or somatosensory cortices. J Neurophysiol 85(3):1078–1087

Malmierca MS, Ryugo DK (2011) Chapter 9: Descending connections of auditory cortex to the midbrain and brain stem. In: Winer JA, Schreier CE (eds) The auditory cortex. Springer, New York, pp 189–208

Marler P, Peters S, Ball GF, Dufty AM Jr, Wingfield JC (1988) The role of sex steroids in the acquisition and production of birdsong. Nature 336(6201):770–772

Matsui R, Tanabe Y, Watanabe D (2012) Avian adeno-associated virus vector efficiently transduces neurons in the embryonic and post-embryonic chicken brain. PLoS One 7(11):e48730

Mittmann W, Wallace DJ, Czubayko U, Herb JT, Schaefer AT, Looger LL, Denk W, Kerr JN (2011) Two-photon calcium imaging of evoked activity from L5 somatosensory neurons in vivo. Nat Neurosci 14(8):1089–1093

Morrison RG, Nottebohm F (1993) Role of a telencephalic nucleus in the delayed song learning of socially isolated zebra finches. J Neurobiol 24(8):1045–1064

Murayama M, Pérez-Garci E, Lüscher HR, Larkum ME (2007) Fiberoptic system for recording dendritic calcium signals in layer 5 neocortical pyramidal cells in freely moving rats. J Neurophysiol 98(3):1791–1805

Nordeen KW, Nordeen EJ (1992) Auditory feedback is necessary for the maintenance of stereotyped song in adult zebra finches. Behav Neural Biol 57(1):58–66

Nottebohm F (1980) Testosterone triggers growth of brain vocal control nuclei in adult female canaries. Brain Res 189(2):429–436

Nottebohm F (1981) A brain for all seasons: cyclical anatomical changes in song control nuclei of the canary brain. Science 214(4527):1368–1370
Nottebohm F, Kelley DB, Paton JA (1982) Connections of vocal control nuclei in the canary telencephalon. J Comp Neurol 207(4):344–357
Ono M, Oliver DL (2014) The balance of excitatory and inhibitory synaptic inputs for coding sound location. J Neurosci 34(10):3779–3792
Polley DB, Steinberg EE, Merzenich MM (2006) Perceptual learning directs auditory cortical map reorganization through top-down influences. J Neurosci 26(18):4970–4982
Rasmusson DD, Clow K, Szerb JC (1992) Frequency-dependent increase in cortical acetylcholine release evoked by stimulation of the nucleus basalis magnocellularis in the rat. Brain Res 594(1):150–154
Roberts TF, Tschida KA, Klein ME, Mooney R (2010) Rapid spine stabilization and synaptic enhancement at the onset of behavioural learning. Nature 463(7283):948–952
Roberts TF, Gobes SM, Murugan M, Ölveczky BP, Mooney R (2012) Motor circuits are required to encode a sensory model for imitative learning. Nat Neurosci 15(10):1454–1459
Saldaña E, Feliciano M, Mugnaini E (1996) Distribution of descending projections from primary auditory neocortex to inferior colliculus mimics the topography of intracollicular projections. J Comp Neurol 371(1):15–40
Scharff C, Nottebohm F (1991) A comparative study of the behavioral deficits following lesions of various parts of the zebra finch song system: implications for vocal learning. J Neurosci 11(9):2896–2913
Schofield BR (2009) Projections to the inferior colliculus from layer VI cells of auditory cortex. Neuroscience 159(1):246–258
Sohrabji F, Nordeen EJ, Nordeen KW (1990) Selective impairment of song learning following lesions of a forebrain nucleus in the juvenile zebra finch. Behav Neural Biol 53(1):51–63
Steriade M, Nuñez A, Amzica F (1993) A novel slow (< 1 Hz) oscillation of neocortical neurons in vivo: depolarizing and hyperpolarizing components. J Neurosci 13(8):3252–3265
Stosiek C, Garaschuk O, Holthoff K, Konnerth A (2003) In vivo two-photon calcium imaging of neuronal networks. Proc Natl Acad Sci U S A 100(12):7319–7324
Suga N, Ma X (2003) Multiparametric cortico-fugal modulation and plasticity in the auditory system. Nat Rev Neurosci 4(10):783–794
Tanaka M, Sun F, Li Y, Mooney R (2018) A mesocortical dopamine circuit enables the cultural transmission of vocal behavior. Nature 563:117–120
Vallentin D, Kosche G, Lipkind D, Long MA (2016) Neural circuits. Inhibition protects acquired song segments during vocal learning in zebra finches. Science 351(6270):267–271
Vu ET, Mazurek ME, Kuo YC (1994) Identification of a forebrain motor programming network for the learned song of zebra finches. J Neurosci 14(11 Pt 2):6924–6934
Wang Y, Brzozowska-Prechtl A, Karten HJ (2010) Laminar and columnar auditory cortex in avian brain. Proc Natl Acad Sci U S A 107(28):12676–12681
Wang Y, Zorio DAR, Karten HJ (2017) Heterogeneous organization and connectivity of the chicken auditory thalamus (Gallus gallus). J Comp Neurol 525(14):3044–3071
Weinberger NM (1995) Dynamic regulation of receptive fields and maps in the adult sensory cortex. Annu Rev Neurosci 18:129–158
Whaling CS, Soha JA, Nelson DA, Lasley B, Marler P (1998) Photoperiod and tutor access affect the process of vocal learning. Anim Behav 56(5):1075–1082
Wild JM, Karten HJ, Frost BJ (1993) Connections of the auditory forebrain in the pigeon (Columba livia). J Comp Neurol 337(1):32–62
Winer JA, Larue DT, Diehl JJ, Hefti BJ (1998) Auditory cortical projections to the cat inferior colliculus. J Comp Neurol 400(2):147–174
Yan J, Ehret G (2002) Corticofugal modulation of midbrain sound processing in the house mouse. Eur J Neurosci 16(1):119–128
Yan W, Suga N (1998) Cortico-fugal modulation of the midbrain frequency map in the bat auditory system. Nat Neurosci 1(1):54–58

Yan J, Zhang Y, Ehret G (2005) Cortico-fugal shaping of frequency tuning curves in the central nucleus of the inferior colliculus of mice. J Neurophysiol 93(1):71–83
Zhang Y, Suga N (2000) Modulation of responses and frequency tuning of thalamic and collicular neurons by cortical activation in mustached bats. J Neurophysiol 84(1):325–333
Zheng W, Knudsen EI (1999) Functional selection of adaptive auditory space map by GABAA-mediated inhibition. Science 284(5416):962–965
Zhou X, Jen PH (2000) Brief and short-term cortico-fugal modulation of subcortical auditory responses in the big brown bat, Eptesicus fuscus. J Neurophysiol 84(6):3083–3087
Zhou X, Jen PH (2005) Cortico-fugal modulation of directional sensitivity in the midbrain of the big brown bat, Eptesicus fuscus. Hear Res 203(1–2):201–215
Zhou X, Panizzutti R, de Villers-Sidani E, Madeira C, Merzenich MM (2011) Natural restoration of critical period plasticity in the juvenile and adult primary auditory cortex. J Neurosci 31(15):5625–5634

MIX
Papier aus verantwortungsvollen Quellen
Paper from responsible sources
FSC® C105338

If you have any concerns about our products,
you can contact us on
ProductSafety@springernature.com

In case Publisher is established outside the EU,
the EU authorized representative is:
Springer Nature Customer Service Center GmbH
Europaplatz 3, 69115 Heidelberg, Germany

Printed by Libri Plureos GmbH
in Hamburg, Germany